BURLEIGH DODDS SCIENCE: INSTANT INSIGHTS

NUMBER 89

Advances in fertilisers and fertiliser technology

Published by Burleigh Dodds Science Publishing Limited
82 High Street, Sawston, Cambridge CB22 3HJ, UK
www.bdspublishing.com

Burleigh Dodds Science Publishing, 1518 Walnut Street, Suite 900, Philadelphia, PA 19102-3406, USA

First published 2023 by Burleigh Dodds Science Publishing Limited

British Library Cataloguing in Publication Data
A catalogue record for this book is available from the British Library

ISBN 978-1-80146-651-6 (Print)
ISBN 978-1-80146-652-3 (ePub)

DOI: 10.19103/9781801466523

Typeset by Deanta Global Publishing Services, Dublin, Ireland

Contents

Series list

Title	Series number
Sweetpotato	01
Fusarium in cereals	02
Vertical farming in horticulture	03
Nutraceuticals in fruit and vegetables	04
Climate change, insect pests and invasive species	05
Metabolic disorders in dairy cattle	06
Mastitis in dairy cattle	07
Heat stress in dairy cattle	08
African swine fever	09
Pesticide residues in agriculture	10
Fruit losses and waste	11
Improving crop nutrient use efficiency	12
Antibiotics in poultry production	13
Bone health in poultry	14
Feather-pecking in poultry	15
Environmental impact of livestock production	16
Sensor technologies in livestock monitoring	17
Improving piglet welfare	18
Crop biofortification	19
Crop rotations	20
Cover crops	21
Plant growth-promoting rhizobacteria	22
Arbuscular mycorrhizal fungi	23
Nematode pests in agriculture	24
Drought-resistant crops	25
Advances in detecting and forecasting crop pests and diseases	26
Mycotoxin detection and control	27
Mite pests in agriculture	28
Supporting cereal production in sub-Saharan Africa	29
Lameness in dairy cattle	30
Infertility and other reproductive disorders in dairy cattle	31
Alternatives to antibiotics in pig production	32
Integrated crop-livestock systems	33
Genetic modification of crops	34

Chapter 1

Spray technologies in precision agriculture

Paul Miller, Silsoe Spray Applications Unit Ltd, UK

1 Introduction

The application of plant protection products plays a key role in the production of most crops. Methods of application depend on the type of crop being treated, with the main ones being:

- Field crops - mainly treated with boom sprayers;
- Bush and tree crops - treated with air-assisted or tunnel-type sprayers;
- Glasshouse and specialist food crops - treated with purpose-designed and -built equipment - often at small scale, manually operated and with handheld delivery systems.

This chapter is concerned only with equipment designed to operate with field crops although many of the principles discussed will relate to other application systems. For field crops grown in Northern European conditions, the costs of plant protection products typically account for between 30% and 50% of the direct costs of production, and so there are important financial incentives in ensuring that such products are applied in a manner that maximises their effectiveness and minimises the quantities applied. Many such products have

http://dx.doi.org/10.19103/AS.2017.0032.09

a toxicological profile that means that it is important to avoid exposure to non-target surfaces and organisms and their use is regulated based partly on the effect that such substances may have on non-target systems. An important performance parameter for any application system therefore relates to the extent to which off-target exposure can be minimised.

When considering the role that spray applications can play within precision agriculture approaches for field crops, there are three scales that are important, namely:

- A whole field/plot area scale where it is important to ensure that the applied dose is effectively controlled and delivered in a manner that will maximise effectiveness and minimise losses to non-target systems;
- A patch spraying scale that involves treating parts of a field area differently. Each patch can be subject to separate operations determined from sensing systems used at the time of application or from a treatment map generated from data collected at other times;
- A plant scale treatment where applications are directed at single plants or small groups of plants with a spot treatment applied to the detected plant(s). Such approaches are usually operated in real time with a sensing system to identify plant(s) to be treated and a delivery system that applies product only to the detected targets and with the minimum contamination of surrounding plants and areas.

Each of the above scales will involve particular characteristics of the application system, but there will be many features that are common to all scales. In this chapter, some general features of spraying equipment for treating field crops that are relevant to precision agriculture are considered before addressing features that are relevant to each of the above scales and reviewing some case studies.

It has been recognised that computer-based control systems are likely to be at the centre of any spray application system operating as part of a precision agriculture environment with key components shown in Fig. 1 concerning:

- Inputs relating to location and time, the physical properties of the liquid to be delivered, condition of the crop and target pest/weed/disease and weather conditions;
- An on-board memory containing details of the machine characteristics including the specifications of the delivery nozzle that will be updated for each batch of treatments requiring a given set-up;
- A loadable memory containing information that will include a treatment map for individual fields;

- A control algorithm for the delivery system including sensors to provide feedback information;
- A means to automatically monitor and record the treatments applied.

2 Features of field crop sprayers for precision agriculture

2.1 Control of delivered dose

For most conventional crop sprayers, the delivered dose is controlled by selecting a nozzle size to deliver a defined flow rate that, at a given operating pressure and nominal forward speed, will apply a set volume of spray liquid per unit area. This is verified in a calibration routine and the machine then operates to apply the set volume of liquid containing a fixed concentration of active product(s) loaded into the machine. Changes in applied dose that would result from changes in the nominal forward speed are accommodated by changes in pressure at the nozzle, and the machine will typically monitor pressure and/or flow rates to give information to the operator as well as a basis for the pressure-based dose control system. However, since flow rate through an orifice (the nozzle) is a function of the square root of pressure, relatively large changes in pressure are required for a given change in flow rate. In practice, this limits the turn-down ratio that can be achieved with conventional dose control systems to some 1.0:1.25 when fitted with conventional nozzles and 1.0:1.40 when fitted with 'extended range/variable pressure' nozzles. This ratio is mainly a function of the spray volume distribution pattern achieved by such nozzles although

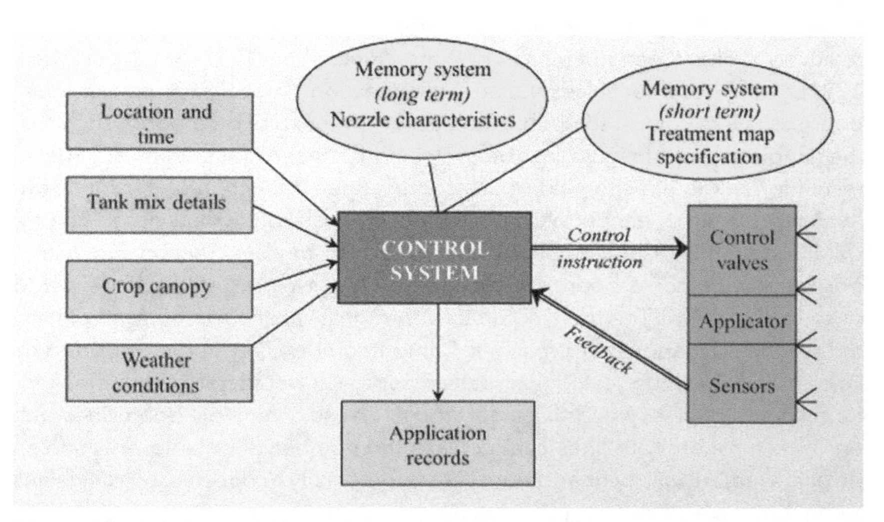

Figure 1 Typical structure of a sensing and control system for a crop sprayer operating as part of a precision agriculture approach. Source: Adapted from Miller (2003a).

it has been recognised that changes to the droplet size distribution with changing pressure may further limit this ratio (Paice et al., 1996). Combellack and Miller (1999) proposed that the definition of turn-down ratio for spatially variable application should be based on a constant droplet size distribution with a tolerance in the volume median diameter (v.m.d.) of ±10%, but there is no evidence that this has been accepted or taken up commercially. Most of the turn-down ratio associated with conventional and extended range flat fan nozzles will probably be used, accommodating variations in spraying speeds such that there is little scope to use pressure control systems to change the applied dose in a precision spraying system. Paice et al. (1996) indicated that a turn-down ratio of 3:1 would be desirable to accommodate forward speed variations, but in practice the use of a set calibrated speed and a defined field layout will mean that turn-down ratios in the order of 1.5:1 are adequate to compensate for variations in forward speed in many situations. However, other approaches have been used to change applied dose levels, and some of these have also been used to extend the speed range over which effective control of the delivered dose can be maintained as well as facilitating variable dose applications. Most dose control systems currently available for precision application change applied dose by changing the volume of a dilute mixture of plant protection products and water such that changes in dose will also result in changes in the applied volume of spray liquid.

2.1.1 On/off control

On/off control is probably the simplest form of dose control that can be used on crop sprayers and this can be implemented using solenoid valves having relatively rapid response times (<10.0 ms - Rietz et al., 1997; Miller and Watt, 1980), giving good spatial resolution in the direction of travel. Most conventional boom sprayers control the output to sections of nozzles on a boom such that the simplest control would involve switching boom sections on or off. Spatial resolution in the direction along the boom is then a function of section width (typically 3.0–6.0 m) although the current trend towards individual nozzle control will improve this resolution, typically down to 0.5 m - see Section 2.2. With a single spray line on a boom, on/off control only gives two levels of delivered dose - full and zero. This then requires a very high level of performance from any detection and mapping approaches. Computer modelling studies reported by Paice et al. (1998) suggested that such an approach would not give stable long-term control of grass weeds in cereal crops. The use of multiple boom lines with on/off control on both lines can increase the number of dose levels that can be achieved but will incur additional cost, complexity and increased difficulties when flushing and cleaning the sprayer.

On/off controls have been used in conjunction with other control strategies in the design of systems for the spatially variable application of plant protection products - for example Gerhards and Oebel (2006) and Paice et al. (1996).

2.1.2 Nozzle clusters (multiple nozzle systems)

Multiple nozzle holders that mount up to four nozzles in a single holder have been developed commercially (Sökefeld, 2010) and used to accommodate changes in forward speed and the requirements for spatially variable application. Nozzles can be switched individually or simultaneously to obtain a wide range of outputs such that turn-down ratios of more than 10:1 can be achieved but with some changes in the droplet size distribution/spray quality. Miller et al. (1997) described the performance of such a system in which a three-nozzle arrangement could achieve a turn-down ratio of 5:1, with the nozzles in each assembly being actuated pneumatically. The sequencing of the nozzle on/off control in each cluster allowed for the time taken to switch off a spray and establish a spray following the opening of the control valves such that gaps in the output delivery were avoided.

The arrangement of using multiple nozzles in a single holder can also be used to adjust the physical characteristics of a delivered spray to match spatially variable requirements. For example, when spraying close to a field boundary, there may be a requirement to minimise the risk of drift even if this increases the risk of reduced efficacy in the performance of the applied products. Miller et al. (1997) suggested that a switching system based on two nozzles would provide an adequate range of dose delivery to accommodate forward speed changes. Therefore, if a four-nozzle assembly were to be fitted with two conventional flat nozzles and two air-induction nozzles, then the conventional nozzles could be used to treat the main part of a field and the air-induction nozzles used close to the field boundary where high levels of drift control are required with the switching done automatically from a field map.

2.1.3 Twin-fluid nozzles and spinning discs

Twin-fluid (or gaseous energy) nozzles use both pressurised air and liquid supplies into the nozzle body and, for a given nozzle configuration, the resulting flow rate and droplet size distribution is a function of the two pressures (Matthews et al., 2014). Studies by Western et al. (1989) confirmed that nozzle output and the droplet size distribution could be controlled independently for relatively low flow rates and that the drift risk with such nozzles was lower than that for conventional nozzles operating at comparable flow rates. Sökefeld (2010) suggested that early commercial designs of twin-fluid nozzle ('Airtec' by Cleanacres Machinery Ltd, Hazleton, Cheltenham, Gloucestershire, GL54 4DX,

UK and 'AirJet' by TeeJet Technologies, Wheaton Facility, 200 W North Avenue, Glendale Heights, Illinois, 60139, USA) were capable of turn-down ratios of 2:1 whilst retaining the same droplet size distribution/spray quality.

A subsequent design of twin-fluid nozzle specifically aimed at use in precision agriculture described by Combellack and Miller (1998, 1999) achieved turn-down ratios of up to 5.5:1 whilst retaining an almost constant droplet size and with an air input requirement that was some 50% lower than that of earlier designs. Wind tunnel studies with this nozzle design also showed that it could operate with levels of drift in the order of 25% of those of conventional nozzles at comparable flow rates.

For spinning disc and cage systems, the droplet size produced is a function of both liquid flow rate and rotational speed, and therefore a spray with a given mean droplet size can be generated over a range of flow rates by adjusting rotational speed. Such a characteristic could be valuable when aiming to vary the dose of chemical delivered but with a defined droplet size distribution/spray quality. Frost (1981) showed that the range of flow rates that could be used to produce a given droplet size increased as the droplet size increased. Heijne (1978) showed that a commercial spinning disc system could achieve a turn-down ratio of 7:1 when mean droplet sizes ranged from 100 to 350 μm. Parkin and Siddique (1990) reported results obtained with a rotary cage atomiser, showing a 10:1 turn-down ratio (flow rates from 2.0 to 20.0 L/min) for mean droplet sizes in the range of 100–180 μm.

2.1.4 Pulse width modulation

Pulse width modulation is a common method used for controlling electrically powered devices by turning the power supply on and off very quickly. Giles and Comino (1990) described a system in which a rapid response solenoid valve was closely coupled to a spray nozzle, with the valve being operated at frequencies in the order of 10 Hz. Flow through the assembly was controlled by the valve duty cycle – the proportion of the time that the valve was held open for each cycle. Results from studies with this system showed that when operating with flat fan nozzles, the system could achieve a turn-down ratio of 3.21:1, with the droplet size increasing as the flow rate was reduced such that v.m.d. increased by some 8.1% across the flow rate range. Giles et al. (1999) used this control approach to develop a system for the precision application of agricultural plant protection products. Trials with this system showed that following a 3:1 step increase in flow rate, the system overshot the set point by approximately 20% but was within 5% of the set point in 5 seconds. Since the system enabled operating pressures to be reduced and flow rate maintained, this enabled the small droplet component within a spray to be reduced – a strategy that could then be used to control drift when operating in sensitive areas such as those close to field boundaries. Recent commercial developments based on this principle have integrated the automatic control of nozzle pressure and duty cycle into a unit that is then compatible with conventional spray rate controllers.

2.1.5 Variable orifice nozzles

Variable orifice nozzle designs have been developed based on different operating principles. Some designs have used materials or a collection of elements arranged to form an orifice configured such that as pressure is fed into the nozzle, the size of the output orifice increases by elastic deformation. The performance of such systems tends to be variable and unpredictable, and there are few performance data relating to such systems in the published literature. An adaptation of this principle for applying liquid fertilisers has been described by Brady et al. (2014) in which a pre-orifice in a stream jet nozzle was covered by a flexible sheath that expands as pressure is applied to the nozzle inlet. Results from tests with the design showed that turn-down ratios in the order of 5:1 could be achieved compared with a value of 1.7:1 for a conventional stream jet nozzle.

A number of alternative designs typically use a spring and/or diaphragm arrangement to adjust the size of a metering orifice within a nozzle by moving a conical shaped plunger positioned above the orifice. As the supply pressure is increased, the conical plunger lifts out of the orifice, increasing the effective orifice area and thus giving a larger increase in flow rate than what would have been achieved through a fixed orifice. Womac and Bui (2002) described and analysed the performance of such an arrangement that used a diaphragm and a control liquid pressure to adjust the position of the metering plunger. They found that a turn-down ratio of 13:1 could be achieved in output flow rates, with the sensitivity to input pressure being related to the control pressure on the diaphragm as expected. For a given control pressure, droplet size increased almost linearly with flow rate (and hence input pressure), with smaller droplet sizes measured at the higher pressure settings (control and line pressures) as expected. The performance of a commercial design of variable orifice nozzle (the 'Varitarget' nozzle) was studied by Daggupati (2007) who found that the design could achieve a turn-down ratio in flow rate of around 3.5:1 when operating at pressures of between 300 and 700 kPa, with commercial data suggesting that turn-down ratios as high as 10:1 can be achieved over a pressure range of 100–600 kPa and depending on the nozzle tip fitted to the assembly but without close control of the droplet size distribution produced. The results presented by Daggupati (2007) confirmed that mean droplet sizes tended to increase with increasing flow rate through this design of nozzle system.

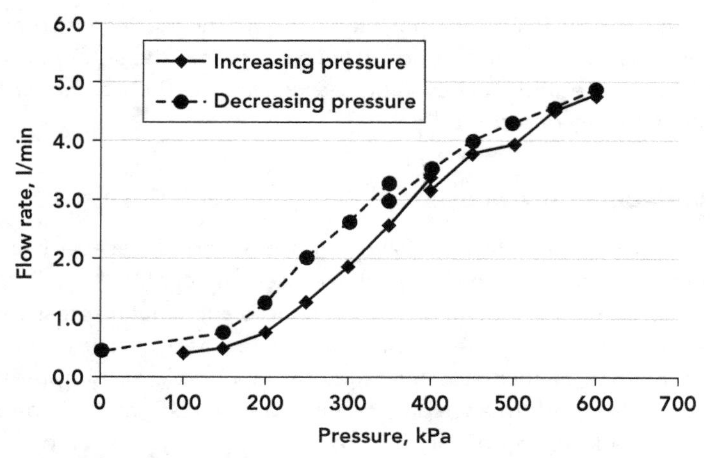

Figure 2 Measured characteristics of a commercial design of variable orifice nozzle showing the wide range of flow rate and the effect of hysteresis.

An important feature of the performance of nozzle systems involving moving parts, such as the 'Varitarget' nozzle, is hysteresis – the difference between a flow rate setting when flow is increasing and that when flow is decreasing. Unpublished data for this type of nozzle design suggest that hysteresis can result in flow variations of up to ±30% of a nominal or mean value – see Fig. 2.

2.1.6 Injection metering systems

Injection metering systems are a potentially attractive method for controlling the dose delivery of a crop sprayer since keeping the diluent (normally water) separate from the formulated products until the time of application can reduce the machine contamination and hence the wash-out and flushing requirements as well as enabling different active ingredients at different doses to be applied in a spatially variable manner. Injection metering also enables the dose of chemical formulation(s) to be varied independently of the applied volume and to target different formulations in a spatially variable manner.

There are two main arrangements for injection metering systems on crop sprayers (Miller, 2003b):

1 With the metering pump(s) positioned upstream of the main system pump. In this case, the metering pump is operating against a relatively low pressure head, and therefore, a wide range of pump types can be used including the peristaltic design. However, since the output capacity of the main pump is greater than the maximum flow rate through all

the nozzles, there needs to be a re-circulation loop such that unused dilute spray liquid is returned to the input side of the main pump. This in turn means that when the system responds to a step change in the demanded output, the response has a ramped characteristic as the effects of the recirculating mixture are accounted for. Because dilute plant protection products are passing through the main pump, there is greater level of machine contamination than is the case when injection takes place downstream of the main pump - see (2) below. Pipework carrying dilute chemicals tends to be relatively long, and there is no opportunity to deliver different mixtures or concentrations to different parts of the boom.

2 With the metering pump(s) positioned downstream of the main pump system. With this arrangement, the metering pump is working to deliver formulated product into spray lines that are at full spraying pressure, and therefore, the metering pump needs a characteristic that is largely independent of the pressure that it must deliver into. This means that piston pumps are commonly used for such applications, but the main pump is operating only with the diluent (water). Injection can take place at any point between the main pump and the nozzles and, although some allowance needs to be made for mixing, experiments have shown that this can generally be achieved in substantially less than 1.0 m length of pipe. Miller and Smith (1992) proposed a system that directly injected formulated products into the nozzle, but it is likely that such an approach would require a metering pump for each nozzle since dividing a metered flow could give inaccurate application if, for example, a nozzle feed line became blocked. This makes such a system relatively expensive and complex. Having long lengths of feed line carrying concentrated formulated products under pressure can give problems with respect to the risk of failure (leakage) and with cleaning and flushing procedures.

Any injection metering system can operate in two possible modes:

1 With variable spray concentration, such that changes in dose are delivered by changing the concentration of formulation(s) in the spray liquid. Changes to the formulations included in the spray liquid can also be accommodated with this approach. This mode of operation has the potential advantage that the nozzle used can be selected to give spray characteristics that closely match target requirements (see Section 2.3) and be used for the range of dose deliveries. However, system response times can then be relatively slow as changes in concentration move through the pipework in the sprayer to the nozzles. Antuniassi et al. (1997) examined the response characteristics of some injection

metering systems and indicated that response times to a step change in demand may be greater than 15 s particularly where injection is on the input side of the main pump. Frost (1990) discussed how plumbing components can be used in practical designs and Paice et al. (1995) showed that response times could be substantially reduced by using small bore pipework and accounting for pressure drops that this may incur via the system control unit.

2 With a constant spray concentration, such that changes in dose are delivered by changing the volume of spray liquid using one of the approaches outlined above. The main use of the injection metering system operating in this mode is to keep the active formulated products separate from the water and hence facilitate reduced residual spray liquid and the need for flushing and cleaning, rather than the control of delivered dose.

All injection metering pumps must be able to operate with the full range of plant protection product formulations. This will involve the ability to resist chemical attack and corrosion and to accommodate a wide range of flow characteristics including viscosity which may, for example, vary with temperature. Frost (1990) described a system in which formulated products are delivered from a metering cylinder by pumping water into the base of the cylinder and with the formulated product separated from the water by a piston. In this arrangement, the metering pump needs only to overcome friction within the system, and the use of a calibrated gear pump acted as both pump and flow meter in a closed loop control arrangement. This system was used in an experimental patch sprayer described by Paice et al. (1995) but has not been developed beyond the commercial prototype stage.

While a number of injection metering systems have been developed commercially, the uptake of such systems has been limited probably due to the following factors identified by a number of authors including Sudduth et al. (1995):

1 The relatively high cost and complexity of such systems.
2 The difficulty in accommodating a wide range of formulations. Most injection metering systems are aimed at using liquid formulations, although systems for granules (but not powders) have been developed, and the wide range of concentrate volumes that constitute a single dose that makes the design of a system that will accommodate the full range very difficult. Metering pumps will typically operate over a 10:1 range of flow rates, while formulations may have full dose volumes ranging from 250 mL/ha to 5.0 L/ha. Many systems therefore have different metering

pump size options to accommodate the range of formulations - again adding to the expense and complexity.

3 The interface with the formulation container, the need in some cases to return any unused formulation back into its original container and problems associated with the cleaning, flushing and priming of delivery lines carrying concentrated formulations.

2.1.7 Other systems

Bjugstad et al. (2012) described a system for control of dose delivery using two tanks - one loaded with dilute formulations at the maximum concentration to be used and the other with clean water. Each tank was connected to a conventional diaphragm pump and the flows from the two tanks were mixed to change the delivered concentration depending on spatial specification (a treatment map) and the speed of the sprayer. Results of tests with an experimental machine showed that the system delivered the required dose within ±5.0% for 95% of the treated area and had the advantage of being relatively insensitive to the physical properties of the formulations.

2.2 Spatial resolution

The resolution at which spatially selective spraying is operated is likely to be a major factor in determining its cost-effectiveness. Two considerations are relevant:

- the spatial distributions of the weed/disease/pest to be treated together with the timescales over which these might change; and
- the characteristics of the delivery system that will influence both the lateral and longitudinal resolution that can be achieved.

Much of the analysis examining the required spatial resolution has focused on the application of herbicides to weed patches, with fewer studies directed at fungicides or insecticides. Grass weed patches in arable fields are relatively stable and persistent, and therefore, mapping such patches as a basis for deriving a treatment map is feasible. Rew et al. (1996, 1997) analysed the effects of spatial resolution when treating patches of grass weeds in arable crops and, as expected, found that the savings in herbicide use reduced as spatial resolution was increased. For example, using a typical strategy for controlling common couch (*Elymus repens*) and increasing the spatial resolution from 2.0 to 4.0 m reduced the herbicide saving from 70% down to 63%. Paice et al. (1996) concluded that, considering the likely increased cost of mapping and treatment at the finer scale, the practical optimum resolution

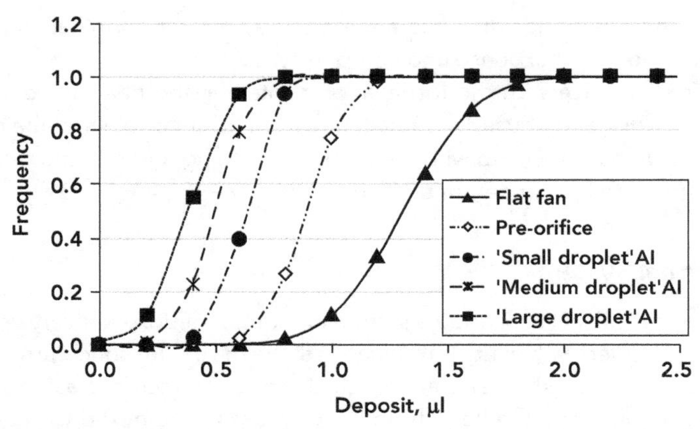

Figure 3 Deposit distributions measured on vertical 1.0-mm diameter rods sprayed with different nozzle designs, all having a flow rate of 0.8 L/min at a pressure of 300 kPa. Source: Adapted from Miller et al. (2010).

was likely to be at least 4.0 m. Miller and Lutman (2008) also recognised the higher costs associated with operating at a finer resolution particularly when the mapping of weed patches was based on manual surveying. It is possible that the potential development of automated weed detection methods may enable mapping and treatment at a small resolution in an economically viable way (Murdoch et al., 2014).

While fungal diseases may develop in patches (Dammer, 2010), currently such disease patches would need to be mapped manually and treated quickly after mapping since, in certain conditions, the disease could spread rapidly. As with weed patches, sensors to automatically detect crop diseases would change the options for applying spatially variable treatments, but such sensors are not yet commercially available. This suggests that a resolution equal to the width of a boom would be adequate for such treatments. Similarly, the spatial resolution to treat pest infestations is unlikely to be at less than the width of the boom.

Most of the options for controlling the dose delivered by crop sprayers outlined in Section 2.1 operate across the full width of a boom. The current trend for increasing boom widths on commercial sprayer designs for operating in field crops (now up to 48 m) means that if spatially variable application were to be implemented using current equipment, the lateral scale of resolution would be very high. While it is technically feasible to envisage separate pressure control or injection metering systems for different boom sections, this substantially increases the cost and complexity of the complete machine, and this has not been attempted commercially to date. On/off control with either single or multiple spray lines can be readily implemented on conventional sprayer designs at a scale corresponding to boom section widths – typically

3.0–6.0 m. Commercial sprayers are now available with individual nozzle control via Controller Area Network (CAN) (e.g. Househam Sprayers Ltd, Woodhall Spa, Lincolnshire, LN10 6YQ, UK), and this approach would facilitate a finer lateral resolution for on/off control and potentially for pulse width modulation systems. Nozzles are commonly spaced at 0.5 m intervals along a boom but arranged to give sprays that fully overlap so that any control strategy at the resolution of the nozzle spacing will not have a sharp cut-off edge.

Longitudinal resolution is a function of the response time of the sprayer and the spraying speed. Systems based only on the use of solenoid valves are therefore likely to have a longitudinal resolution approximately equivalent to boom section widths (circa 1.0 s at 4.0 m/s). Those with pressure control systems would be approximately equivalent to the boom length, and some injection metering systems would be much longer but predictable so that a control system could look ahead with map-based systems.

2.3 Matching physical characteristics of sprays to target requirements

While it is likely that most precision spray applications would involve variations in dose, it has been shown that different target geometries retain different proportions of an emitted spray depending on the spray quality/droplet size distribution generated by the nozzle. Results from laboratory studies measuring deposits on idealised 1.0 mm stainless steel rods mounted vertically showed large differences between different nozzle designs producing different droplet size distributions – see Fig. 3.

The results in Fig. 3 have been linked with results from field trials that show higher levels of control of grass weeds in cereals crops when treated with finer sprays having a smaller mean droplet size distribution (Butler Ellis et al., 2008). However, the use of finer sprays will incur a greater risk of drift (see Section 2.4) such that it may be appropriate to use a finer spray only when treating a patch of grass weeds. There is therefore scope to consider spatially variable applications using nozzles with characteristics that are directly matched to the spray target particularly when using multiple nozzle systems as discussed in Section 2.1.2 although such approaches have not been widely developed to date.

2.4 Minimising drift and exposure of systems outside the treatment area

The control of spray drift is now an important aspect relating to the operation of field crop sprayers with implications for managing the exposure of non-target organisms and surfaces. Using buffer zones, both within the cropped

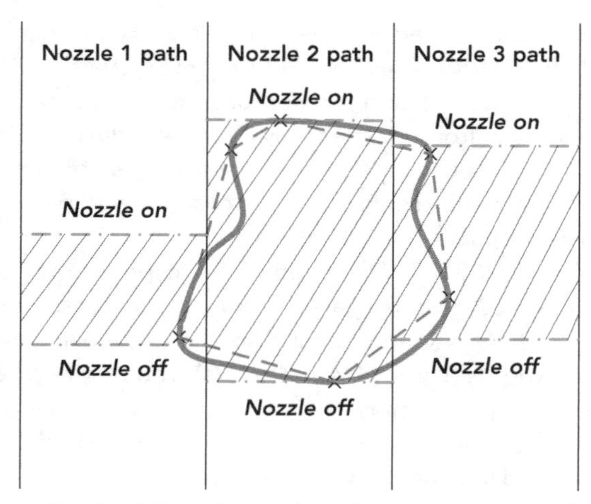

Figure 4 Diagram showing how a detected weed shape was mapped to the control of spray nozzles.

Figure 5 An example of an analysed image showing the centre of a carrot crop bed (blue crosses), crop rows (as green lines) and detected volunteer potatoes bounded by polygons (in red).

area and at the edge of the crop, is now an established method for controlling the exposure to pesticides of non-target systems such as surface water and non-target plants in many countries. Concerns relating to toxicity levels and the sensitivity of non-target systems to pesticide exposure in recent years have led to proposals to increase the widths of such buffer zones, enabling the approval of products where the degree of risk reduction offered by a smaller buffer zone would not be sufficient (Bryon and Hamey, 2008). The use of relatively large buffer zones is not popular with farmers. One way of

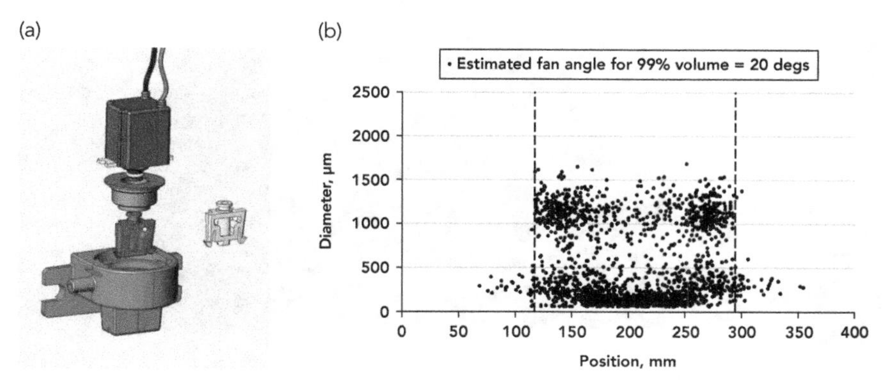

Figure 6 (a) Left: nozzle and solenoid valve cartridge assembly showing two options for the fitted spray tip. (b) Right: droplet size and spatial distribution pattern for the nozzle operating at 75 kPa pressure as measured with an imaging system (see Tuck et al., 1997) showing large droplets with a bi-modal size distribution and a relatively sharp cut-off at the edge of the pattern.

Figure 7 Pattern achieved from the fluidic nozzle supplied with water and tracer dye at 0.75 bar running right to left at a height of 0.5 m and a speed of 2.10 m/s.

enabling the widths of buffer zones to be reduced is to use nozzles/spraying arrangements that have been shown to reduce drift when operating close to field boundaries – for example as in the LERAP scheme in the UK (Anon., 2001). The widths of buffer zones, the extent by which they can be reduced and the widths of field area close to the boundary that must be sprayed with the drift-reducing method depend on the details of the scheme and the product(s) to be applied. Farmers may be reluctant to use the drift-reducing application arrangements over the whole field area because of a risk to efficacy. Mapping the regions close to field boundaries with the specifications relating to buffer zone implementation allows compliance to

(a)

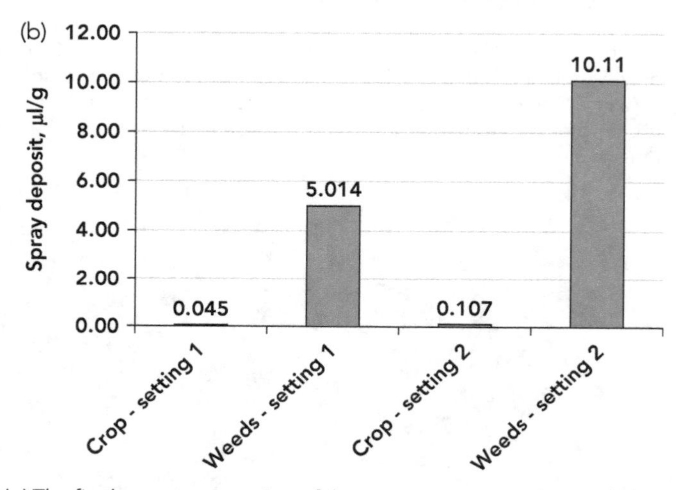

Figure 8 (a) The final prototype version of the spot spraying system working in a crop of leeks and (b) the results of measurements of spray deposits on target weeds and crop plants within 150 mm radius of a treated plant for two settings of the machine.

be achieved automatically with an added advantage that such compliance can be recorded.

3 Case study 1: designing and developing a system for spot treatment of volunteer potatoes

The control of volunteer potatoes in crops such as onions, leeks and carrots is an important issue influencing the yield, quality and ease of harvesting of such crops. Traditionally the control of such weeds has been by repeated overall spraying of selective herbicides, but the revision of European Regulations EU 91/414 relating to the registration of active substances for plant protection has meant that many of the formulated products used for such applications are no longer available. The options to use selective herbicides for the control of weeds such as volunteer potatoes are now very limited. Spot treatment with a total herbicide was identified as a possible alternative to overall spraying based on the technology available (e.g. Hague and Tillett, 2001) and the fit with the agronomic requirements for the vegetable crops considered. The work to develop the spot spraying system involved:

1. The development of a weed detection algorithm (not discussed in detail in this chapter);
2. The development of an application system that would give adequate deposits on the target weed with the minimum potential contamination of surrounding crop plants;
3. The design and construction of an experimental prototype machine;
4. Laboratory experiments and field trials to assess the performance of system components and the complete machine.

The weed detection algorithm identified the plan outline of a detected weed and positioned this with respect to a spray boom and nozzles that would apply the total herbicide (Miller et al., 2012a) - see Fig. 4. The process of identifying the weed outlines and associated spray nozzle schedules was repeated at a rate of 30 Hz in successive images as they passed through the field of view of the camera system (Fig. 5) to refine the calculation of the area to be treated. Provision was made to spray a proportion of this plan area as a variable that could be adjusted in the field depending on the relative risks of damage to neighbouring crop plants and the control of the weed. This was determined by assessing whether nozzles at the edge of the area to be sprayed could remain off while still achieving the required percentage of the treatment area. If this was the case, then the switching on and off for nozzles passing over the central part of the target was eroded equally until the preset proportion of weed plan area to be treated was reached.

The specification for the nozzles concerned the uniform application of herbicide sprays based on existing formulation types, to areas less than 100 mm square when travelling at a speed of 5.0 km/h with the minimum of drift and splash outside of the treated area. This involved the rapid establishment and cut-off for a spray in a total cycle time of 0.025 s. Nozzle spacing was to be matched to crop rows but would be in the order of 100 mm. A review of potential nozzle systems meeting this specification identified two candidate designs: a fluidic type that generated an oscillating liquid stream from the nozzle by a passive arrangement of feed channels within the nozzle body and an oscillating hypodermic syringe nozzle in which the needle was mounted on a vibrating arm and excited by an electrical vibrator. An evaluation of these two options was conducted (Miller et al., 2012b), and while both were shown to technically meet the specification, the fluidic design was selected for a

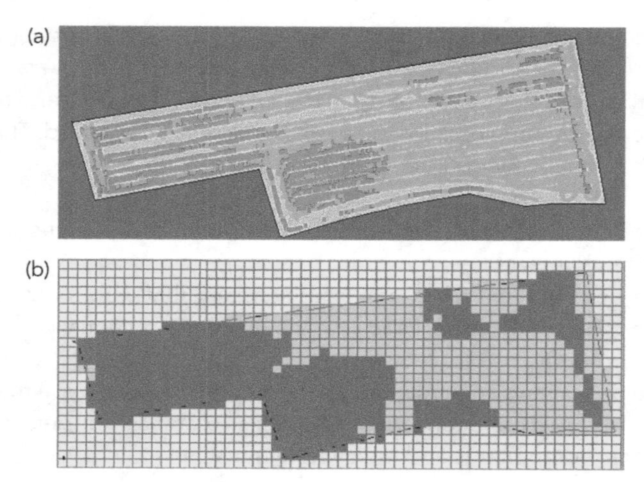

Figure 9 (a) Weed patches as mapped visually from a combine harvester; (b) treatment map derived from the weed patch map.

Figure 10 Experimental patch sprayer operating in a crop of peas.

prototype machine based on commercial considerations. Further development of this nozzle design was then undertaken in conjunction with Hypro EU Ltd, Station Road, Longstanton, Cambridge, CB24 3DS, UK, to produce a final design (the 'Alternator' nozzle - see Fig. 6a) that incorporated a small latching solenoid matched to the nozzle body and the ability to change spray tips within the nozzle body using a cartridge system. The performance of the nozzle assembly was assessed in terms of the droplet size (Fig. 6b), velocity and spray volume distribution patterns (Fig. 7) generated both with a static nozzle in still air and with a moving nozzle in wind tunnel conditions (Miller et al., 2012b).

The performance of the unit was evaluated in a range of crop and field conditions, assessing the levels of weed control achieved, damage to crop plants and the deposition of sprays both on target weeds and the surrounding crop plants - see Fig. 8. Detailed performance assessments were conducted when spraying mugwort (*Artemisia vulgaris*) in a crop of leeks with two machine settings: one spraying the central 40% of a weed image (conservative strategy aimed at minimising the risk of crop damage) and the other spraying 80% of a weed image (riskier strategy aimed at a high weed kill). Results showed that deposits on targeted weeds were more than two orders of magnitude greater than those on surrounding crop plants (Fig. 8b), with deposits for the two settings directly as expected. Overall results from a series of field trials in a range of crop, weed and weather conditions showed that the unit could achieve levels of weed control that were at least competitive with all alternative approaches and with commercially acceptable levels of crop damage (Miller et al., 2012a).

It was recognised that a key factor leading to the high levels of weed kill and low levels of crop damage recorded with the system was that the large weeds targeted retained the relatively large droplets effectively. Provision was made with the nozzle cartridge assemblies to fit nozzles, giving much finer spray qualities for treating patches of small grass weeds in a range of crop situations. As a result of the experimental work and prototype development, a commercial machine was produced and offered for sale by Garford Farm Machinery Ltd, Hards Lane, Frognall, Deeping St. James, Peterborough, PE6 8RP, UK, who partnered the project work throughout.

4 Case study 2: a patch spraying system for applying herbicides to field crops

When considering the design and development of a patch spraying system that would provide a first-level step for the commercial introduction of such a system, the approaches as demonstrated and described by Paice et al. (1995) based on a combination of injection metering, multiple spray lines, small bore delivery pipes and on/off control were regarded as too complex and expensive for such an introductory system. It was therefore decided to use a multiple nozzle system (see Section 2.1.2) using three conventional nozzles to deliver flow rates over a 5:1 range at a defined spray quality. An important objective for the development was the treatment of grass weed patches in cereal crops and therefore the ability to maintain a fine/medium spray quality was important - see Section 2.3.

A summary specification for the system was defined as follows (Miller, 2003b; Miller et al., 1997):

- An accuracy of delivered dose to better than ±5.0% of target;
- A spatial resolution in the transverse direction to be 4.0 m or greater with a response time of the dose control system of less than 1.0 s such that resolution in the longitudinal direction was of a similar magnitude (depending on spraying speed);

- An ability to operate with a wide range of commercially available formulations.

The control system for the mounted sprayer was linked to a unit in the tractor that supported a field location system based on satellite navigation (GPS) and which also carried the treatment map specifying the doses to be applied in different parts of the field. Weed patch mapping was by manual observations recorded using the same basic unit as mounted on the tractor either using an all-terrain scouting vehicle or with the unit mounted in a combine harvester and weed presence/absence recorded using a push button system. Recorded weed patch maps (Fig. 9a) were transformed into treatment maps (Fig. 9b) using software that added buffers around identified weed patches to account for:

- seed movement since visual mapping was undertaken in one season to provide treatment maps that would be used in the following season;
- any positioning errors during either mapping or treatment;
- the response time characteristics of the dose control system.

The spray delivery system initially used three nozzles at each position along the boom with the switching of nozzles and control of spraying pressure implemented using a compressed air arrangement powered with an air compressor driven from the pump shaft on the sprayer. Experience from field trials with the unit indicated that for many applications, an adequate turn-down ratio could be achieved using a two-nozzle arrangement and this configuration was used in the later trials with the system.

The results from full-scale field trials with the system (Fig. 10) showed that the sprayer could deliver the dose to the specification, but that the development of treatment maps based on visual mapping of weed patches was time-consuming, expensive and prone to errors.

5 Conclusion

Much of the discussion in this chapter has been set against the background of treating crops grown on relatively wide areas including the example of the spot treatment of large weeds in vegetable crops grown in rows, and this has strongly influenced the technologies that have been discussed. There is a considerable body of work that has examined the application of chemical and non-chemical treatments to individual plants using versions of autonomous robots, and such systems are likely to have applications in high-value crops grown in small areas. The effectiveness of all plant protection products is very influenced by the timeliness of application, and hence, work rates are important. Some schedules for applying such products recommend that treatments are to be applied in less than 3 days from the time of assessing the need for the treatment. The types of system included in this review are generally capable of operating in such timescales.

Many of the developments of targeted spraying systems took place in the late 1990s or early 2000s with a focus for much of the work directed at the use of herbicides for weed control because of the evidence of patchiness and the speed of propagation of weeds. The performance of systems has been shown to meet specifications particularly as far as the spatially variable delivery of sprays is concerned. However, the commercial uptake of such systems has been relatively slow. In 2008, Miller and Lutman cited the following reasons for this limited uptake:

- The economic case for adoption is not clear with the costs of equipment and mapping not readily recovered particularly for high dose/low dose strategies;
- Patch spraying is regarded as risky with high risks particularly associated with errors in weed (patch) detection;
- A lack of field data to show that the risks associated with patch spraying are manageable;
- A reluctance by farmers and agronomists to spend time and effort manually mapping weed patches recognising that research investigating automated systems for weed detection could change this in the future;
- No current regulatory incentive to reduce herbicide use even though Government (and European) policy is to minimise use.

The current situation is much as it was in 2008. Some more recent studies have aimed at quantifying the performance and advantages of systems based on the principles outlined in this chapter. For example, work by Zaman et al. (2011) considered the development of a spot sprayer using conventional narrow spray angled nozzles coupled to solenoid valves as part of a system based on the real-time detection of weeds using height discrimination and ultrasonic detectors. Results from this work demonstrated the feasibility of spot application in blueberry crops. González-de-Soto et al. (2016) evaluated the performance of a robotised patch sprayer that used an injection metering system and conventional nozzles as the basis for spray delivery control and concluded that such a system could achieve treatment and control of more than 99.5% of detected weeds with substantial savings in herbicide use. This study involved assumptions relating to idealised weed detection and evaluated the performance of the spray delivery in both laboratory and field conditions. A technical-economic analysis by Tona et al. (2017) suggested that while on-off and canopy adjusted application systems could be cost-effective over conventional approaches, more complex approaches could probably not be justified assuming the likely costs of units required to deliver such treatments. There have been commercial developments over the last decade that have refined the design of a number of the systems described in this chapter particularly to reduce costs and improve performance and reliability. There are now indications that some of these systems are being involved in developments with realistic commercial objectives and potential outcomes.

Work to automatically detect weed patches has made progress (see e.g. Murdoch et al., 2014; Weis and Sökefeld, 2010), and is considered elsewhere in this book, but there is currently no commercially available system operating to do so. Problems relating to herbicide resistance for some grass weed species have become more serious, and now some farmers are mapping the heaviest infestations of blackgrass patches in cereal crops with a view to applying a total herbicide to the patch destroying both weed and crop but preventing seed return to the soil in the patch.

Many sprayers designed for operation in field crops now have a computer-based control system with links to a field location system using satellite navigation methods and a map-based recording system. Such systems are already delivering improved precision in spray application and will provide the platform for further improvements in the future both with and without links to spatial variability.

6 Future trends

While current crop spraying systems can get data relating to the field in which they are operating from downloaded maps or using field location and mapping systems, all of the data relating to the products that are being applied are currently input manually from the product label or use settings/configurations that the operator derives from information on the product label. If the sprayer control system could have information relating to the product to be applied, then this would enable decisions relating to optimised application to be made automatically particularly if inputs from other sensors relating to the crop and weather conditions are also available - see Fig. 1. The potential for reading product labels on the sprayer and automatically recording the quantities and details of formulations loaded into the machine has been investigated and demonstrated (Watts et al., 2003; Peets et al., 2009), but, to date, these systems have not been developed commercially. The particular advantages of such an approach would relate to:

- An improved automated management of drift risk accounting for local rules and conditions at the time of spraying;
- The potential to account for both crop and weather conditions at the time of spraying - sensors are already available to characterise the crop

canopy, but further research is required to relate crop canopy conditions to application parameters to optimise responses;
- The ability to automatically generate records that fully meet existing and possible future requirements.

It is also likely that new sensing systems for both weeds and diseases will be developed in the future (as discussed elsewhere in this book), and this will have an important influence on the potential for both spatial and temporally variable applications. Work on distinguishing weeds from crop continues to be a major technical subject for study and has been reviewed by a number of authors (e.g. Weis and Sökefeld, 2010). The detection of grass weed patches in cereal crops has recently made progress and is now close to commercial exploitation (see Murdoch et al., 2014 and commercial literature relating to 'eyeweed'). The situation relating to both diseases and insect pests is more challenging. Even with improved sensor technologies (West et al., 2010; Martinelli et al., 2015), it is likely that there will be few patchy diseases that can be detected in time for spatially variable spray treatments to be effective (West et al., 2010).

While spatial variability is an important component of precision spray application, it is also important to deliver the appropriate dose to the appropriate target to maximise biological control. The interface between application systems, decision support tools and improved sensing systems is likely to be the basis for key developments in the future.

There continues to be considerable research and commercial development interest in the platforms from which precision sprayers might operate. Autonomous vehicles have been developed and used experimentally but are not yet a commercial reality. Applications from unmanned aerial platforms (drones) are also being investigated for specialist conditions where the alternative would be manual application and where the limited payload of such aerial systems can be managed effectively.

7 Where to look for further information

Further information on the design and performance of agricultural spraying systems can be found in journals such as *Biosystems Engineering*, *Crop Protection* and *Precision Agriculture*. Details of nozzle design and performance are reported in *Atomization and Sprays*. The key conference relating to Pesticide Application is now that run biannually by the Association of Applied Biologists as 'International Advances in Pesticide Application' and with proceedings published in *Aspects of Applied Biology*. Key conference papers may also be reported in the proceedings of The American Society for Agriculture and Biological Engineers (ASABE).

8 References

Anon. (2001). Local Environmental Risk Assessment for Pesticides (LERAP). Horizontal Boom Sprayers, A step-by-step guide to reducing aquatic buffer zones in the arable sector. Pesticides Safety Directorate, Department for Environment, Food and Rural Affairs, London, UK.

Antuniassi, U. R., Miller, P. C. H. and Paice, M. E. R. (1997). Dynamic and steady-state dose response of some chemical injection metering systems. *Proceedings of the Brighton Crop Protection Conference - Weeds*, 17-20 November 1997. The British Crop Protection Council, Surrey, UK, pp. 587-692.

Bjugstad, N., Ensby, T., Holth, E. H., Moen, O. K. and Thorstensen, H. F. (2012). FlexiDose - a spraying system developed for variable pesticide dosage without using pesticide injection methods. *Aspects of Applied Biology, International Advances in Pesticide Application*, 144, 181-8.

Brady, M. J., Goddard, R. S. and Kately, S. J. (2014). A novel device for the variable rate application of liquid fertilisers 'Trident'. *Aspects of Applied Biology, International Advances in Pesticide Application*, 122, 411-13.

Butler Ellis, M. C., Miller, P. C. H. and Orson, J. H. (2008). Minimising drift while maintaining efficacy - the role of air-induction nozzles. *Aspects of Applied Biology, International Advances in Pesticide Application*, 84, 59-66.

Byron, N. and Hamey, P. (2008). Setting unsprayed buffer zones in the UK. *Aspects of Applied Biology, International Advances in Pesticide Application*, 84, 123-30.

Combellack, J. H. and Miller, P. C. H. (1998). Does the technology exist to efficiently and effectively patch spray weeds? In: *Precision Weed Management in Crops and Pastures*. R. W. Medd and J. E. Prately (Eds). Proceedings of a workshop 5-6 May 1990, Wagga Wagga. Published for CRC for Weed Management Systems, University of Adelaide, South Australia, Australia, pp. 55-61.

Combellack, J. H. and Miller, P. C. H. (1999). The development of a twin-fluid nozzle for precision agriculture. *Proceedings of the 1999 Brighton Conference - Weeds*, 15-18 November 1999. The British Crop Protection Council, Surrey, UK, pp. 473-80.

Daggupati, N. P. (2007). Assessment of the Varitarget nozzle for variable rate application of liquid crop protection products. Master of Science Thesis, Kansas State University, Manhattan, KS.

Dammer, K.-H. (2010). Variable rate application of fungicides. In: *Precision Crop Protection - The Challenge and the Use of Heterogeneity*. E.-C. Oerke, R. Gerhards, G. Menz and R. A. Sikora (Eds). Springer, Dordrecht, Heidelberg, London and New York, pp. 349-63.

Frost, A. R. (1981). Rotary atomisation in the ligament formation mode. *Journal of Agricultural Engineering Research*, 26, 63-78.

Frost, A. R. (1990). A pesticide injection metering system for use on agricultural spraying machines. *Journal of Agricultural Engineering Research*, 46, 55-70.

Gerhards, R. and Oebel, H. (2006). Practical experiences with a system for site-specific weed control in arable crops using real-time image analysis and GPS-controlled patch spraying. *Weed Research*, 46, 185-93.

Giles, D. K. and Comino, J. A. (1990). Droplet size and spray pattern characteristics of an electronic flow controller for spray nozzles. *Journal of Agricultural Engineering Research*, 47, 249-67.

Giles, D. K., Stone, M. L. and Dieball, K. (1999). Distributed network system for control of spray droplet size and application rate for precision chemical application. *Precision Agriculture'99*. J. V. Stafford (Ed.). Proceedings of the 2nd European Conference on Precision Agriculture, Sheffield Academic Press, Sheffield, UK, pp. 857–66.

González-de-Soto, M., Emmi, L., Perez-Ruiz, M., Aguera, J. and Gonzalez-de-Santos, P. (2016). Autonomous systems for precise spraying – Evaluation of a robotised patch sprayer. *Biosystems Engineering*, 146, 165–82.

Hague, T. and Tillett, N. D. (2001). A bandpass filter approach to crop row location and tracking. *Mechatronics*, 11(1), 1–12.

Heijne, C. G. (1978). A study of the effects of disc speed and flow rate on the performance of the Micron 'Battleship'. *Proceedings of the British Crop Protection Conference – Weeds*, 20–23 November 1978. The British Crop Protection Council, Surrey, UK, pp. 20–3.

Martinelli, F., Scalenghe, R., Davino, S., Panno, S., Scuderi, G., Ruisi, P., Villa, P., Stroppiana, D., Boschetti, M., Goulart, L. R., Davis, C. E. and Dandekar, A. M. (2015). Advanced methods of plant disease detection. A review. *Agronomy for Sustainable Development*, 35(1), 1–25.

Matthews, G. A., Bateman, R. and Miller, P. C. H. (2014). *Pesticide Application Methods*. 4th edn. John Wiley and Sons Ltd, West Sussex, UK.

Miller, P. C. H. (2003a). The current and future role of application in improving pesticide use. *Proceedings of the BCPC International Congress Crop Science and Technology*, 10–12 November 2003. The British Crop Protection Council, Surrey, UK, pp. 247–54.

Miller, P. C. H. (2003b). Handling and dose control. In: *Optimising Pesticide Use*. M. Wilson (Ed.). John Wiley and Son Ltd, West Sussex, UK, pp. 45–73.

Miller, P. C. H. and Combellack, J. H. (1997). The performance of an air/liquid nozzle system suitable for applying herbicides in a spatially selective manner. In: *Precision Agriculture'97*. J. V. Stafford (Ed.). Proceedings of the First European Conference on Precision Agriculture, Volume 1. BIOS Scientific Publishers, Oxford, UK, pp. 651–60.

Miller, P. C. H. and Lutman, P. J. W. (2008). A review of the factors influencing the technical feasibility and potential commercial uptake of the patch spraying of herbicides in arable crops. *Aspects of Applied Biology, International Advances in Pesticide Application*, 84, 265–72.

Miller, M. S. and Smith, D. B. (1992). A direct nozzle injection controlled spray boom. *Transactions of the ASAE*, 35(3), 781–5.

Miller, P. C. H. and Watt, B. A. (1980). The use of a high speed photographic technique to examine the response time characteristics of a solenoid spray valve. *Journal of Agricultural Engineering Research*, 25, 217–20.

Miller, P. C. H., Paice, M. E. R. and Ganderton, A. D. (1997). Methods of controlling sprayer output for spatially variable herbicide application. *Proceedings, Brighton Crop Protection Conference – Weeds*. The British Crop Protection Council, Surrey, UK, pp. 641–4.

Miller, P. C. H., Butler Ellis, M. C., Bateman, R., Lane, A. G., O'Sullivan, C. M., Tuck, C. R. and Robinson, T. H. (2010). Deposit distributions on targets with different geometries and treated with a range of spray characteristics. *Aspects of Applied Biology, International Advances in Pesticide Application*, 99, 241–8.

Miller, P. C. H., Tillett, N. D., Hague, A. and Lane, A. G. (2012a). The development and field evaluation of a system for the spot treatment of volunteer potatoes in vegetable crops. *Aspects of Applied Biology, International Advances in Pesticide Application*, 114, 113–20.

Miller, P. C. H., Tillett, N. D., Swan, T., Tuck, C. R. and Lane, A. G. (2012b). The development and evaluation of nozzle systems for use in targeted spot spraying applications. *Aspects of Applied Biology, International Advances in Pesticide Application*, 114, 159–66.

Murdoch, A., Flint, C., Pilgrim, R., De La Warr, P., Camp J, Knight, B., Lutman, P., Magri, B., Miller, P., Robinson, T., Sanford, S. and Walters, N. (2014). eyeWeed: Automating mapping of black-grass (*Alopecurus myosuroides*) for more precise applications of pre- and post-emergence herbicides and detecting potential herbicide resistance. *Aspects of Applied Biology, Crop Production in Southern Britain: Precision Decisions for Profitable Cropping*, 127, 151–8.

Paice, M. E. R., Miller, P. C. H. and Bodle, J. (1995). An experimental machine for evaluating spatially selective herbicide application. *Journal of Agricultural Engineering Research*, 60, 107–16.

Paice, M. E. R., Miller, P. C. H. and Day, W. (1996). Control requirements for spatially selective herbicide sprayers. *Computers and Electronics in Agriculture*, 14, 163–77.

Paice, M. E. R., Day, W., Rew, L. J. and Howard, A. (1998). A stochastic simulation model for evaluating the concept of patch spraying. *Weed Research*, 38, 373–8.

Parkin, C. S. and Siddique, H. A. (1990). Measurements of the drop spectra from rotary cage aerial atomizers. *Crop Protection*, 9, 33–8.

Peets, S., Gasparin, C. P., Blackburn, D. W. K. and Godwin, R. J. (2009). RFID tags for identifying and verifying agrochemicals in food traceability systems. *Precision Agriculture*, 10(5), 382–94.

Rew, L. J., Cussans, G. W., Mugglestone, M. A. and Miller, P. C. H. (1996). A technique for surveying spatial distribution of weeds and the potential reduction in herbicide use from patch spraying (*Elymus repens* L.) common couch, in cereal fields. *Weed Research*, 36, 283–92.

Rew, L. J., Miller, P. C. H. and Paice, M. E. R. (1997). The importance of patch mapping resolution for sprayer control. *Aspects of Applied Biology, Optimising Pesticide Applications*, 48, 49–56.

Rietz, S., Palyi, B., Ganzelmeier, H. and Laszlo, A. (1997). Performance of electronic controls for field sprayers. *Journal of Agricultural Engineering Research*, 68, 399–407.

Sökefeld, M. (2010). Variable rate technology for herbicide application. In: *Precision Crop Protection – the challenge and the use of Heterogeneity*. E.-C. Oerke, R. Gerhards, G. Menz and R. A. Sikora (Eds). Springer, Dordrecht, Heidelberg, London and New York, pp. 335–47.

Sudduth, K., Borgelt, S. C. and Hou, J. (1995). Performance of a chemical injection sprayer system. *Transactions of the ASAE*, 11(3), 343–8.

Tona, E., Calcante, A. and Oberti, R. (2017). The profitability of precision spraying on speciality crops: A technical-economic analysis of protection equipment at increasing technical levels. *Precision Agriculture* online (*in Press*).

Tuck, C. R., Butler Ellis, M. C. and Miller, P. C. H. (1997). Techniques for measuring droplet size and velocity distributions in agricultural sprays. *Crop Protection*, 16(7), 619–28.

Watts, A. J., Miller, P. C. H. and Godwin, R. J. (2003). Automatically recording sprayer inputs to improve traceability and control. *Proceedings, The BCPC International*

Congress - Crop Science and Technology, 1. The British Crop Protection Council, Surrey, UK, pp. 323–8.

Weis, M. and Sökefeld, M. (2010). Detection and identification of weeds. In: *Precision Crop Protection - The Challenge and the Use of Heterogeneity*. E.-C. Oerke, R. Gerhards, G. Menz and R. A. Sikora (Eds). Springer, Dordrecht, Heidelberg, London and New York, pp. 119–34.

West, J. S., Bravo, C., Oberti, R., Moshou, D., Ramon, H. and McCartney, H. A. (2010). Detection of fungal diseases optically and pathogen inoculum by air sampling. In: *Precision Crop Protection - The Challenge and the Use of Heterogeneity*. E.-C. Oerke, R. Gerhards, G. Menz and R. A. Sikora (Eds). Springer, Dordrecht, Heidelberg, London and New York, pp. 135–49.

Western, N. M., Hislop, E. C., Herrington, P. J. and Jones, E. I. (1989). Comparative drift measurements for BCPC reference hydraulic nozzles and for an Airtec Twin-Fluid nozzle under controlled conditions. *Proceedings, The BCPC Conference* - Weeds, 12–15 November 1989. The British Crop Protection Council, Surrey, UK, pp. 641–8.

Womac, A. R. and Bui, Q. D. (2002). Design and tests of a variable flow fan nozzle. *Transactions of the ASAE*, 45(2), 287–95.

Zaman, Q. U., Esau, T. J., Schumann, A. W., Percival, D. C., Chang, Y. K., Read, S. M. and Farooque, A. A. (2011). Development of a prototype automated variable rate sprayer for real-time spot-application of agrochemicals in wild blueberry fields. *Computers and Electronics in Agriculture*, 76, 175–82.

Chapter 2

Site-specific nutrient management systems

Dan S. Long, USDA-ARS, USA

1 Introduction

Farm management has been generally defined as the science of organizing and controlling the resources of a farm such that profitability is maximized (Castle et al., 1972; Nix, 1979). Dillon (1980) succinctly defined farm management as *the process by which resources and situations are manipulated by the farm manager in trying, with less than full information, to achieve his goals*. This statement recognizes the importance of *the dynamic nature of the farm system and its environment, and the uncertain nature of the farmer's decisions, thereby implying attempted rather than sure achievement of objectives*.

Such is the case of site-specific nutrient management in which growers attempt to account for crop nutrient demand that varies in space and time within farm fields. However, ability to do so is largely limited by an incomplete understanding of the causes and magnitude of the variability underlying crop fertilizer response.

Farm management systems are created through the establishment of inputs, processes and outputs that are intended to maximize yield and profit (Fig. 1). By analogy, this concept applies to nutrient management systems as well in which inputs result from the physical environment (e.g. precipitation, temperature and growing season length) and farm management (e.g. labour, machinery, fertilizer, herbicide and other materials). Processes are actions within the farm that are directed towards the production of outputs and include but are not limited to image acquisition, field mapping, zone delineation, soil sampling and fertilizing. Outputs include grain and biomass, dollar returns and improved soil, water and air quality but they can also be negative such as greenhouse gas emissions, herbicide-resistant weeds and soil erosion.

http://dx.doi.org/10.19103/AS.2017.0032.14
Published by Burleigh Dodds Science Publishing, 2019.

Processes are carried out by farmers in a repeating series of steps that begins with inventory of spatial data and their recording in a geographical information system (GIS) followed by geographic analysis and planning, and application of plans through global navigation satellite systems (GNSS)-guided field implements (Fig. 2, Nielsen et al., 1996). Final results are monitored and compared with projected economic and environmental outcomes and recorded for use in the next cycle. As an aside, this 'precision farming cycle' is a variation of the Japanese Plan-Do-Check-Act cycle that has its origin with the Shewhart Cycle (Shewhart, 1939) and Deming Wheel (Deming, 1950) for quality control in the United States (U.S.) manufacturing sector (Fig. 3). The purpose of a cyclical series of actions is to include the development, testing and implementation of changes that will result in improvement to the system of management.

Site-specific nutrient management systems were created to improve nutrient use efficiency and balance trade-offs between profitability and environmental concerns. Fertilizers are prescribed to match crop demand for nutrients that are known to vary in space and time within farm fields. This approach embraces the 4-element 'right product, right rate, right place and right time' advanced by the International Plant Nutrition Institute (Johnston and Bruulselma, 2014). Though understanding of the environmental dynamics affecting site-specific nutrient management may be imperfect, a cornerstone of such management is knowledge evolving from the information acquired, which improves over time with each iteration of the precision farming cycle.

Environmental factors that influence crop production across the conterminous U.S. are terrain, climate, soil and plant available water (Baker and Capel, 2011). It is the combination of these four environmental factors that also determines the development of a specific nutrient management system in a certain area. Nutrient management systems differ in the information utilized and the way they operate to accomplish their common purpose. The objective of this chapter is to illustrate site-specific nutrient management with examples of systems adapted for sub-humid and semi-arid environments. After a general review of valid processes that are presently used to collect information for site-specific nutrient management, there follows examples where the achievements of different systems are described. The final section describes further developmental needs.

2 Processes to inform site-specific nutrient management

2.1 Grid sampling

In the 1990s, systematic grid point sampling of soil and laboratory analysis of resulting cores were used in the U.S. Midwest as a primary means of obtaining discrete values of soil test nitrogen (N), phosphorus (P), potassium

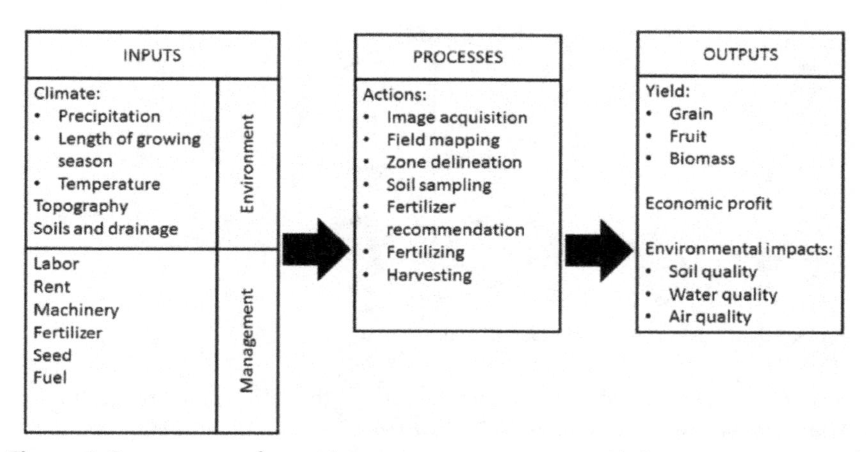

Figure 1 Components of a nutrient management system with inputs, processes and outputs.

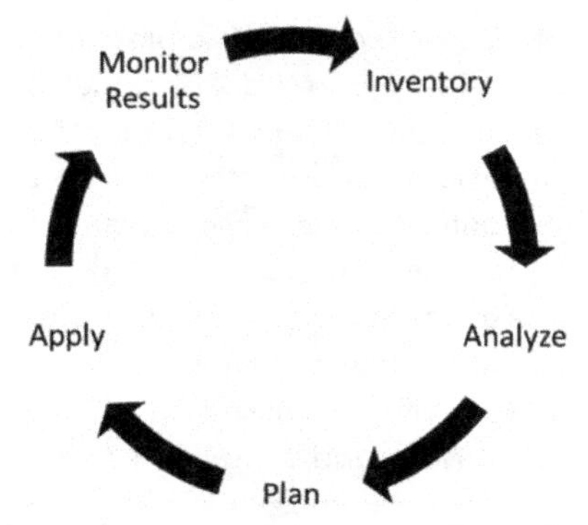

Figure 2 Precision farming cycle represented as a repeating series of actions or processes.

(K) and other soil variables at geographically distributed points within fields (Wollenhaupt et al., 1994). Mathematical interpolation methods such as inverse distance weighting and kriging were then used to estimate values for rasterized maps, which were useful for spatial overlay operations in a GIS. Because of the strong role of topography on the spatial distribution of soil properties, cokriging with high resolution and readily available digital terrain data has been shown to be potentially useful for interpolation of a limited number of soil test P and K values, and to increase the efficiency of grid sampling (Kozar et al., 2002). Nonetheless, contour maps made from interpolated grid data showed that spatial variability in nutrients did not conform to soil survey maps and that nutrients were highly variable within map units. Grid sampling methods that ignored soil and landscape units

Figure 3 Japanese plan, do, check and action cycle.

were then promoted with a 1-ha cell size giving the best spatial resolution for least cost (Mallarino and Wittry, 2001).

Soil testing for immobile nutrients such as P and K is common in many production systems and site-specific management of these nutrients may involve grid sampling. For instance, the grain yield responses of corn and soybean to variable-rate and fixed-rate P fertilization were compared over 4 years in six farm fields in Iowa (Bermudez and Mallarino, 2007). Soil test P sampled on a dense grid within 6–12 ha experimental areas served as a basis for varying dry phosphate fertilizer with commercial spreading equipment. While there was no significant difference in yield between the two management systems, variable-rate P placement showed potential for reducing excess P loss to the environment as a result of reducing the P application rate to high-testing field areas. In addition, variable-rate placement applied 12.4% less P fertilizer and reduced within-field variability in soil test P.

Today, many farmers regard grid sampling as too expensive for routine collection and analysis of soil samples, but it remains common for spatial characterization of topsoil levels of P and K. In contrast, residual NO_3-N levels are highly unstable in soils due to microbial respiration and loss by leaching, and thus the frequent sampling needed for spatial characterization of fields will be cost-prohibitive. Citing greater cost-effectiveness, many growers and their advisers have now adopted the 'zone sampling' approach whereby smaller sub-regions in a field, or management zones, are delineated and sampled separately as described below.

2.2 Constructing management zones

Within-field crop variability is fundamentally the product of interaction between many ecosystem factors including topography, soils, climate, pests, animals and management. However, a cursory examination of a farm field from above is sufficient to show patterns of crop growth that tend to be grouped into different levels forming more or less definite zones. The spatial crop variability is not random, but is distributed in a pattern over the landscape. This natural phenomenon is well illustrated in traditional use of published soil surveys depicting areas of farmland that are expected to be relatively homogeneous (Mausbach and Wilding, 1991) and aerial photographs showing tonal differences in soil moisture and crop growth within fields (Milfred and Kiefer, 1976).

Management zones are defined as sub-regions of a field with homogeneous yield-limiting factors for which a single rate of a specific crop input is appropriate (Doerge, 2000) and are established based on ability to consistently capture field patterns from year to year (Mallarino and Wittry, 2001). By dividing a field into zones and sampling them individually, zone sampling captures spatial nutrient variation while reducing sampling cost. One soil testing laboratory (Midwest Laboratories, Inc., Omaha, NE, USA) recommends that the area represented by each sample be no more than 8 ha with each sample a composite of 15–25 cores within a 6.1 m radius. One of the first studies published on precision agriculture proposed the use of soil testing by soil map unit, which differed in N, P and K (Carr et al., 1991). When the Global Positioning System (GPS) allowed for mapping of soil test values, it became apparent that soil map units could be highly variable in soil nutrient values (Mallarino and Wittry, 2001). General soil maps are not recommended for delineation of N management zones because they seldom contain detailed, site-specific information (Franzen et al., 2002).

One approach to identifying zones is by measurement and mapping of the yield-determining factors at a site (King et al., 2005). Contact electrode and non-contact inductance sensors both provide a rapid, low-cost means for obtaining continuous measurements of apparent electrical conductivity (EC_a) that vary in fields with changes in clay content and mineralogy, porosity, cation exchange capacity, compaction, depth to indurated layer, water content, salinity level and other soil physical properties (Corwin and Lesch, 2003). Spatial patterns in EC_a have been correlated to yield, topsoil depth, and soil physical and chemical properties on certain soils in the U.S. Midwest (Kitchen et al., 1999, 2003). Johnson et al. (2003) showed that management zones derived from EC_a maps were useful for characterizing spatial variation of wheat and corn yields. Farmers are intimately familiar with their fields yet a map of soil EC_a was found to be more effective in identifying homogeneous sub-regions within fields than farmer interpretation of a bare soil image (Fleming et al., 2004). King et al. (2005) concluded that the method can provide useful information for the provisional delineation of soil type boundaries and crop management zones, and that expert site assessment and knowledge were necessary to confirm their existence. Kitchen et al. (2005) stated further that an EC_a map can be a good indicator of underlying soil properties; however, temporal variability must be considered when implementing this approach due to changes in crop response with environmental conditions.

A second approach to identifying zones utilizes the spatial variation in crop yield at a site. Growers are increasingly using yield monitors to acquire site-specific yield data in their fields. In the U.S. between 1996 and 2012, 5–30% of soybean, 8–34% of corn and 2–13% of spring wheat areas were mapped using a yield monitor (Schimmelpfennig, 2016). Though a single-year map may have limited value for supporting management decisions over extended time periods (Ping and Dobermann, 2005), multi-year yield data can be used to produce temporal stability maps showing high or low yielding areas that had not changed from year to year as well as unstable areas that had been highly changeable (Blackmore, 2000). Yield stability maps were used to identify homogeneous zones within fields for site-specific management of P, K and lime (Flowers et al., 2005). Taylor et al. (2001) analysed 3–7 years of yield monitor data for cornfields in Kansas and found the ability of a single year's yield map to predict yield in the following year to vary substantially. Sequential yield maps were useful for deriving a yield goal map only if the correlation among years was consistently high ($r \geq 0.70$).

Research has found that a single source document is often insufficient for guiding site-specific applications (McCann et al., 1996; Franzen et al., 1998). Multiple data layers can provide more information for subdividing fields into management zones than can be achieved with one layer of information. For example, Anderson-Cook et al. (2002) determined that soil EC_a readings and crop yield maps together could separate soil types with 90% accuracy for variable-rate fertilizer management. Hornung et al. (2006) found that a combination of bare soil imagery, topography and farmer experience more effectively defined zones of different yield productivity than a combination of bare soil imagery, soil organic matter, cation exchange capacity, soil texture and previous season's yield map. Schepers et al. (2004) used principal components analysis and unsupervised classification of principal component scores to aggregate variability in soil brightness, elevation and electrical conductivity into four management zones. Though these zones adequately characterized spatial variation in soil pH, soil EC_a, soil test P and soil organic matter, they were less consistent in characterizing variation in crop yields.

The use of a numerical technique (e.g. principle components analysis, unsupervised classification etc.) for the integration of multiple data to produce more useful information than any individual data source is known as data fusion (Liggins et al., 2001). Clustering is a data fusion technique that has been widely used to group data points into homogeneous groups. A number of different algorithms and techniques have been applied to spatial data including crop yield (Stafford et al., 1999); soil EC, elevation and slope (Fridgen et al., 2000); and multi-year cotton yield (Boydell and McBratney, 2002) for the purpose of creating management zones. Milne et al. (2012) created compact zones by applying smoothed fuzzy classification to multi-year yield data but resulting zones had a similar N response suggesting no benefit from fertilizing each zone separately. A soil map of the same field also distinguished compact zones that differed in crop N response such that it could have been more profitable to manage by soil type.

The Management Zone Analyst (MZA) software uses the fuzzy c-means unsupervised clustering algorithm to assign field observations into like classes, or management zones, and determines the appropriate number of zones (Fridgen et al., 2004). Farid et al. (2015) applied the MZA software to point values of percent sand, percent clay, elevation, soil N and soil EC sampled within a wheat field in Pakistan comprising a complex of alluvial soils. Performance indices were available to indicate that the field should be divided into four management zones that significantly differed in grain yield. Other available software for classification of management zones and automatic determination of number include ZoneMap (Zhang et al., 2010) and EZZone (Lowrance et al., 2016). Both tools are web-based and provide for interpolation of point data to a common grid. MZA does not have this feature and instead requires that data be gridded before import.

Management zones are designed to capture within-field variability in plant nutrients, crop yield or crop fertilizer response with the best technique explaining the most variance. In Montana, Long et al. (1995) found that 67% of the variance in final yield could be explained by management zones derived from a mid-season aerial photograph of ripening differences in wheat. In South Dakota, Chang et al. (2003) investigated the impact of different zone mapping techniques on soil NO_3-N and Olsen P sampling variability. Grid cell sampling (3.5 ha) yielded the greatest reduction in sampling variance below that of the whole field followed by management zone classification by GIS analysis of electrical conductivity, elevation, aspect and distance information; cluster analysis of electrical conductivity, elevation and aspect; and 1:12 000 scale soil survey maps. Variance was further reduced when areas historically impacted by homesteads or animals were sampled separately. Later, Chang et al. (2004) showed that variability in university N and P recommendations were best reduced using multiple year yield monitor data to establish yield goals followed by grid cell sampling to reduce N and P sampling error.

Principles of fertilization for zone-based nutrient management are based on sampling soils at or near planting and analysing them in the laboratory for N, P, K and S contents. For N, soil test values together with expected yield and the unit requirement for yield are then used to derive fertilizer recommendations based on university recommendation methods. Accuracy of zone-based N management to target crop N needs depends on correctness of these N fertilizer recommendations. Unfortunately, this approach provides little information on future levels of plant available N that are released over the growing season from mineralization of soil organic matter and crop residues (van Es et al., 2007).

As up to 50% of crop N uptake is supplied by soil (Kramer et al., 2002), it is necessary to predict N mineralization from organic sources to synchronize the N supply with crop N demand and prevent overfertilization with N (Wade et al., 2016). Leaching can also be a major loss mechanism for NO_3-N, particularly in humid environments where infiltration exceeds runoff and nitrate levels exceed crop N demand. Furthermore, loss of NO_3-N by denitrification occurs when soils are saturated with water for prolonged periods. Consequently, economically optimum N rates (EONR) are highly variable within fields and between years at the same site (Dhital and Raun, 2016), yield goals are unpredictable (Raun et al., 2017) and they correlate poorly with recommended N rates (Scharf, 2015). Spatio-temporal variability in environmental factors affecting these N cycling processes is why responsive in-season crop sensing may be warranted.

2.3 In-season crop sensing

Beginning in the mid-late 1990s, researchers in the central Great Plains of the U.S. investigated the feasibility of active optical sensing of the crop canopy for determining crop N status for in-season N management (Stone et al., 1996; Solie et al., 1996; Sembiring et al., 1998). The rationale for in-season N management is that the information on crop N status is obtained when crop N demand is greatest and applying N at this time would improve N use efficiency. Spectral properties of the crop canopy had been found useful in predicting crop foliar N content (Curran, 1989). Work with the Minolta SPAD 502 chlorophyll meter had also indicated that analysis of differences in light reflectance from plant leaves would be useful for predicting N fertilizer needs for corn in humid regions (Schepers et al., 1992). Yield losses from N deficiency were noted to occur when chlorophyll meter readings were <95% of those for well-fertilized corn and this value was proposed as a cut-off point for applying supplemental N. Other studies with corn showed that spatial patterns in N deficiency and crop response to added N could be detected and mapped through analysis of aerial colour imagery (Blackmer et al., 1996). Deficiencies of N, seen as a lighter green colour on aerial imagery, result in leaves having less chlorophyll, less photosynthesis and increased reflectance of radiation normally absorbed as red light during photosynthesis.

The Normalized Difference Vegetation Index (NDVI), computed from radiance in the Red (630–690 nm) and NIR (760–900 nm) wavebands, is directly related to aboveground green biomass (Rouse et al., 1974). It is indirectly related to chlorophyll concentration through the negative effect of N deficiency on crop cover and crop biomass. Researchers in Oklahoma developed an in-season N management strategy that calculates N fertilizer rates as a function of an in-season estimate of grain yield and the likely yield response to additional N fertilizer (Raun et al., 2002). Crop canopy reflectance in the red and NIR converted to NDVI is continuously obtained with a downward-pointing sensor mounted to a boom on a high-wheel applicator. An in-season estimate of yield (INSEY) in an unfertilized reference area is derived by dividing NDVI by the number of days from planting to sensing. The projected N uptake for potential yield with added N is determined from the INSEY times an N-response index computed as the ratio of the NDVI of an N-rich reference and NDVI from elsewhere in the field. The fertilizer N recommendation is then computed by subtracting the projected N uptake for the potential yield in the N limiting sensed area from INSEY, and dividing by a factor (between 0.6 and 0.7) representing the expected uptake efficiency to obtain the in-season top dress N rate (Shanahan et al., 2008).

Much of the N that is applied to corn in the U.S. Midwest is lost to the environment. Part of the problem can be traced to the failure of soil-based diagnostic tools to accurately determine plant N demand. In Missouri, Scharf and Lory (2009) found that the N rate needed to economically optimize corn yield varies widely from field to field and within fields. Spatial variability in the N supplying power of soil appeared to be a major determinant of crop response to applied N and thus the optimal N rate. Field experiments showed that corn colour could explain 53–77% of the variance in corn N demand compared to potential yield or conventional soil tests that only explained <25% of variance. Therefore, the use of active light sensors was proposed for detecting corn colour based on the rationale that N-sufficient corn is darker green and taller than N-deficient corn that is lighter green and shorter. As with the chlorophyll meter, the crop's reflectance from any place in a field is compared with the reflectance from a high-N reference strip. The greater the difference in reflectance between under-fertilized corn and the N-sufficient reference, the more N that is needed. Before crop canopy colour can reliably reflect soil N fertility, corn must be more than 30 cm tall (>V6 growth stage 3 weeks after emergence, Abendroth et al., 2011) to ensure adequate root development and absorption of plant nutrients within the soil root zone.

One problem with the N-rich strip concept is that excessive levels of N can increase the potential for sulphur (S) deficiency in corn resulting in a general yellowing of new leaves, artificially low reference values and low fertilizer N recommendations. In eastern Kansas, Holland and Schepers (2010) developed a 'turn-key' approach that eliminated the need for N-rich reference strips. Their 'virtual reference strip' method relies upon an optical crop sensor with sensitivity in the 'red-edge' (680–730 nm) waveband, which is highly dependent upon chlorophyll concentration (Horler et al., 1982). Operating from a high-clearance applicator, the boom-mounted sensor acquires crop reflectance values over a full range in plant vigour from a small area of a field. Somewhere within this area there is enough N to produce full yield such that the 95th percentile of a histogram of the sensor's crop reflectance values can serve as a reference value of N sufficiency. Accordingly, field areas are identified with adequately fertilized plants, which minimizes the potential for sulphur (S) imbalances caused by excess N availability.

2.4 Grain quality sensing

Crop canopy reflectance is related to plant N through the indirect effect of N fertility on crop biomass. This concept works well where plant available water is plentiful and N is limiting but not in semi-arid environments where dryland crop growth is largely determined by water. Other problems with optical sensing include too little biomass and cover in early season, absence of variation in crop colour and excessive wetness for wheeled applicators. Fertilizer issues include poor nutrient uptake efficiency due to lack of rainfall to move the N into the root zone, stranding of broadcast granular N at the surface in dry soil and N losses due to volatilization. Rainfall is not required for late season foliar applications of liquid N fertilizer but the potential for increasing grain protein concentration (GPC) of hard wheat is difficult to predict.

As an alternative to in-season crop sensing, rugged optical spectrometers with visible-NIR sensitivity are commercially available for installation on combine harvesters to measure and map the GPC in farm fields. Thousands of protein points can be mapped across fields and are used to identify areas of N nutrition adequacy based on the causal relationship between GPC and soil N fertility (Long et al., 2005). Introduced in 2001, the ProSpectra™ Grain Analyzer (Textron Systems) was the first reflectance spectrometer developed for operational use on a combine (von Rosenberg, et al., 2000). Field tests of this instrument on an operating combine (<5.0 g protein kg^{-1} grain) were comparable with a laboratory-grade spectrometer (Long et al., 2008). Other on-combine instruments over the past 10 years included the CropScan 2000H (NIR Technology Systems, Bankstown, NSW, Australia) and AccuHarvest Grain Analyzer (Zeltex, Inc., Gaithersburg, MA, USA) but are no longer sold and have been superseded by the newer CropScan 3000H On-Header Analyzer (Next Instruments, Condell Park, NSW, Australia).

Use of GPC as a diagnostic tool in evaluating whether N was sufficient or deficient for wheat yield once a critical protein level has been identified for a particular region was first proposed by Goos et al. (1982). Critical protein levels have been found to vary by wheat class, cultivar and region. Critical protein levels can be useful for evaluating the potential yield loss to N deficiency. For example, Glenn et al. (1985) found a 9.8% yield reduction in 'Stephens' soft white winter wheat for each grain protein percentage below the critical GPC. Selles and Zentner (2001) concluded that the critical GPC as an indicator of N deficiency works well when water is not limiting but is unreliable when grain yields are negatively impacted by water stress because high protein can also occur under drought. Long et al. (2017) determined that soft white spring wheat under severe water stress with protein below the critical level did not guarantee that yields were compromised by lack of N, but rather N was sufficient above this level.

Together with yield monitor data, site-specific protein data can be used to calculate the amount of N to apply to replace that exported from the field in the harvested crop of the current year. In turn, management zones can be easily constructed to guide N fertilization of the following crop and prevent over-application that stimulates vegetative growth before the onset of drought, lowers yield and elevates protein. A map of N removal (NR) can be easily computed from maps of grain yield and protein and the following equation: $NR = (GY \times GPC) \div 0.17$, where GY is grain yield in kg ha^{-1}, GPC is expressed as a fraction and 0.17 is the fractional part of protein that is N (Engel et al., 1999; Long et al., 2000). Fertilizing with such a map would involve maintaining plant available N for the following crop by replacing only what was removed by the crop in the previous season. If yield of the harvested crop was already limited by poor nutrition, as indicated by GPC below the critical protein level, the N in deficient areas of fields can be increased to raise protein to the critical GPC level with knowledge of the amount of fertilizer N equivalent (FNE) to a unit change in protein. Current estimates of FNE vary considerably and are reported to be 14–22 kg N ha^{-1} for HRS wheat (<3000 kg ha^{-1} yield) in northern Montana (Engel et al., 1999) and 59–82 kg N ha^{-1} for HRS wheat and 54–67 kg N ha^{-1} for SWS wheat in eastern Oregon (Long et al., 2017).

3 Regional perspectives

3.1 Case studies from sub-humid U.S. Midwest

Current cropland yields in the U.S. Midwest are near their maximum production potential largely because of good soils and mild weather but an abundance of plant available nutrients and water is also responsible for the high productivity. Extended periods of drought have become infrequent since 1965 (Andresen et al., 2012) and many states within the region are now experiencing dramatic increases in the number of extreme precipitation events (Frankson et al., 2017). Sufficient N, P and K are necessary to help crops achieve their relatively high yield potential. However, leaching and other loss mechanisms not only result in nutrient runoff to the environment but also prevent nutrients from reaching plants. Therefore, a major focus of site-specific nutrient management systems in the U.S. Corn Belt is on keeping nutrients in the soil and plant available to the plant.

Published by Burleigh Dodds Science Publishing, 2019.

3.1.1 Mid-season nitrogen application to winter wheat

Responsive in-season approaches to site-specific N management started in Oklahoma with the system developed by Raun et al. (2002) that coupled plant sensing with an N fertilizer optimization algorithm that predicted yield at mid-season and the potential N removed in grain. Solie et al. (2012) improved the algorithm by introducing a generalized model that adjusts the yield calibration curve for biomass accumulation that changes with growth stage of both corn and wheat. The algorithm was further developed by Bushong et al. (2016a) who altered the proxy for crop growth (i.e. growing degree days) to count only if soil moisture was adequate for growth. In another report, Bushong et al. (2016b) evaluated current plant-sensor algorithms for predicting a difference in grain yield potential with and without added N fertilizer and concluded that they could effectively predict final yield and agronomic optimum N rate of winter wheat. Despite this progress, producers have been slow to adopt optical plant sensing. Previously, Biermacher et al. (2009) showed that advantages of in-season application of UAN are outweighed by the low cost of pre-plant application of anhydrous ammonia such that both methods are equal in economic performance. Boyer et al. (2011) found no significant difference in net returns between optical reflectance-based N management and conventional management, which clearly indicated that growers would not adopt the technology.

Few reports have been found in the literature in which the Oklahoma system was shown to improve N use efficiency and reduce pollution of ground and surface waters. Recently, however, Cao et al. (2017) documented the potential benefits of responsive, in-season N management to food security, environmental pollution and climate change in China - a country where mismanagement of N fertilizer is widespread and N recovery efficiency is poor. Impacts of precision N management were investigated in a small plot experiment in a rotation of winter wheat and summer maize. Nitrogen was applied as urea to the following six treatments:

1. Zero N fertilizer as a control
2. Conventional grower practice of applying 150 kg N ha^{-1} before planting of both crops and 150 kg ha^{-1} at stem elongation of wheat and V10 stage of maize (5 weeks after emergence, Abendroth et al., 2011)
3. University recommendation of applying 60 kg ha^{-1} at planting and 120 kg ha^{-1} at stem elongation of wheat and V10 of maize
4. 60 kg ha^{-1} at planting and topdressing of 190 kg ha^{-1} at stem elongation of wheat and 127 kg ha^{-1} at V10 of maize based on visual examination of six ramp calibration plots

5 Soil test-based application of 60 kg ha^{-1} at planting and topdressing with allowance for soil NO_3-N content at rates of 148 kg ha^{-1} at stem elongation of wheat and 187 at V10 of maize
6 Application of 60 kg ha^{-1} at planting followed by sensor-based in-season N topdress application of 123 kg N ha^{-1} at stem elongation of wheat and 115 kg ha^{-1} at V10 of maize.

The sixth treatment associated with a plant sensing-based system was consistently better for both crops than any other treatment. Compared to conventional practices, applied N was reduced by <62% thus indicating increased N use efficiencies of <123%, decreased apparent total N losses <81% and lowered emissions of greenhouse gases and N losses of <68% with no significant loss in crop yield.

3.1.2 Responsive nitrogen management in corn

The conventional practice of applying the same rate of N fertilizer to whole fields ignores inherent within-field variability in the economically optimal N fertilizer rate (EONR), known as the point where the last increment of N returns a grain yield increase large enough to pay for that N (Sawyer et al., 2006). Lory and Scharf (2003) found that yield goal-based N recommendations exceeded EONR by 90 kg ha^{-1} on average. Citing evidence that N mineralization of soil organic matter varies widely from site-year to site-year, they postulated that variability in EONR was due to spatial variation in soil N supply rather than crop yield and N demand. Leaf colour, as measured with the SPAD chlorophyll meter, is a good predictor of EONR (Scharf et al., 2006). In Missouri, this finding is the basis for crop canopy sensor-based N management and the translation of active canopy sensor data into in-season N rate decisions (Kitchen et al., 2010).

Between 2004 and 2008, Scharf et al. (2011) compared the agronomic, economic and environmental performance of sensor-based variable-rate N and producer-based constant-rate N applications in 55 cornfields across central Missouri to give a wide range in production environments. Crop reflectance sensors were installed on the front of fertilizer application equipment and positioned directly over the cornrow at a height of 50 cm above the canopy. Each of two active light sensors: Crop Circle 210 (Holland Scientific, Lincoln, Nebraska, USA) and GreenSeeker (NTech Industries, Ukiah, California, USA) was used to direct variable-rate sidedress N applications to corn at V6 to V16 growth stages. Because of a diversity of equipment in use across the farms, methods of placement and forms of N fertilizer included dribbled or injected liquid UAN, injected anhydrous ammonia and broadcast urea.

Published by Burleigh Dodds Science Publishing, 2019.

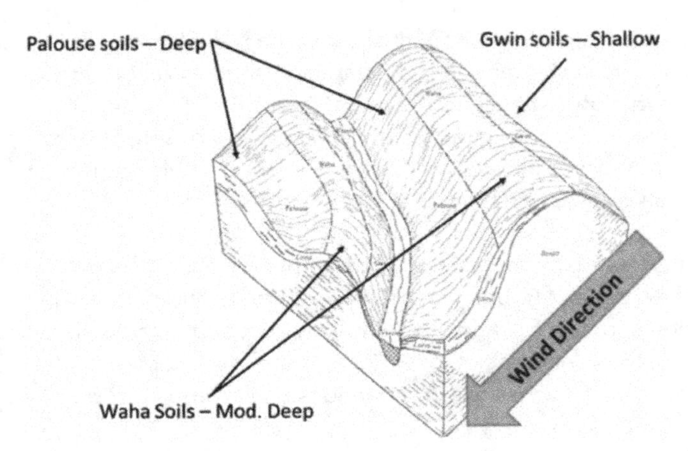

Figure 4 Ridge and valley pattern of soil map units on Herb March farm near Milton-Freewater, Oregon, USA. Block diagram adapted from USDA-SCS Soil Survey of Umatilla County Area, Oregon.

Fixed N sidedress rate chosen by the farmer and variable-rate N sidedress rate were applied in linear treatment plots across each field. During fertilizer application, the sensor-based N rate varied as a function of visible/NIR reflectance values in the field divided by visible/NIR values from a high-N reference area. After harvest, the mean yields of the two placement techniques were compared. By reducing N use and increasing yield, profitability was increased by an average of \$42 ha^{-1} over all fields. Profit was greatest when the sensor-recommended N rate was more N or substantially less N than producer-chosen rates. Sensor use reduced the total amount of applied N by 8% and reduced surplus N, or N applied in excess of that removed in grain at harvest, by 24–26%. These results confirmed that responsive, in-season N placement performed better than conventional uniform methods used by producers to establish sidedress N rates.

3.2 Case studies from semi-arid U.S. West

Plant available water is the most important factor that limits dryland crop yields in semi-arid environments. If the water supply is insufficient, plants will be incapable of utilizing all of the N and other nutrients that are available for growth and development. Therefore, site-specific nutrient management systems in the semi-arid west have largely focused on characterizing spatial patterns in fields that are associated with water-related attributes. Specialized variable-rate equipment and sampling costs are an important consideration as gross returns are often <\$200 ha^{-1} because of the relatively low yields.

Published by Burleigh Dodds Science Publishing, 2019.

Figure 5 ProSpectra Grain Quality Analyzer mounted to the grain bin filling auger of Herb March's John Deere 6622 combine harvester.

Table 1 Yield and protein concentration in 18 wheat fields varying in area, elevation and rainfall across the Herb March farm near Milton-Freewater, Oregon, USA

Field	Area (ha)	Yield (kg ha^{-1})	Protein (g kg^{-1})	Elevation (m)	Rainfall (mm)
North Spence	61.7	3205	114	365	375
Bowman	41.0	5026	97	395	375
NE Garriott	80.4	4616	107	426	400
S Kelly Winter Wht	62.8	4522	108	441	400
S Kinman	74.1	4482	105	456	400
W Shumway	19.2	4824	101	456	400
S Kelly Spring Wht	62.8	1277	131	456	400
Roburn	7.2	4354	107	471	400
240	49.8	5073	107	486	375
Lawrence	232.3	4401	114	502	425
Home	67.5	4502	115	502	450
400	74.6	5254	104	532	450
Troyer (Split)	28.0	5792	102	532	450
LK (Split)	76.5	4925	110	532	475
Old Ranch	62.3	4166	112	608	475
Storey-Waldner	39.7	5066	112	638	475
West Jon	50.2	4636	108	684	525
U.K./West Wiljon	91.4	5207	111	699	525

3.2.1 Terrain-based nitrogen management in hard red spring wheat

In glacial till landscapes of the northern Great Plains, slope position effects on soil water-holding capacity may occur when tillage erosion causes soil loss and reduced rooting depth in upper slopes and soil deposition and increased rooting depth in lower slopes (Moulin et al., 1994). Dividing hill slopes into upper, middle and lower slopes has been used as a method to construct management zones in this region (Beckie et al., 1997; Nolan et al., 1995; Pennock and Corre, 2001). In northern Montana, Long et al. (2014) compared the hard red spring wheat yields and economic returns of variable-rate and uniform-rate N management in eight farm fields between 1994 and 2004. Digital elevation models and terrain analysis were used to divide fields into upper, south-facing middle slopes, north-facing middle slopes and lower slopes. Nitrogen recommendations for management zones were based on soil N testing and expected yields. Nitrogen fertilizer was applied as a replicated series of incremental N rates in treatment strips across each field. Fertilizer was either broadcast from a truck-mounted applicator or banded below the seed with an air seeder. Strips were cut with a production-size combine equipped with a mass flow yield monitor. At the same time, grain was hand sampled from the combine's grain bin filling auger in 20–30 s intervals during travel within each strip. These samples were later analysed for GPC in the laboratory and used to establish the dollar value of the wheat.

Mean grain protein and yield were similar between variable-rate and uniform-rate N management. Yield differences were <223 kg ha^{-1} and averaged only 18 kg ha^{-1}. Grain yield did not differ significantly among N rates within management zones. In seven of the eight sites, net returns from variable-rate N management were up to U.S.$27.97 ha^{-1} less than from uniform N management and were not profitable if Environmental Quality Incentive Program payments of U.S.$6.36 ha^{-1} were considered as part of net income. Little evidence existed that variable-rate N management based on terrain-based management zones and predetermined N recommendations improves grain yields and profits or reduces N use in water-limited, summer fallow systems of northern Montana. The reasons were traced to modest yield gains and N savings, nominal response to applied N, and inability to define management zones and optimal N rates.

3.2.2 Soil depth-based nitrogen management in soft white winter wheat

In the Pacific Northwest, soft white winter wheat is grown as an export crop to the Asian Pacific market that prefers grain with low protein concentration (<105 g kg^{-1}) for production of noodles and pastries. With help from USDA scientists, Herb March of Couse Creek Ranches (Milton-Freewater, Oregon,

Published by Burleigh Dodds Science Publishing, 2019.

USA) implemented precision N practices as a means of improving the protein levels in a winter wheat-chemical fallow production system. The 2439 ha farm is situated on the foothills of the Blue Mountains with elevation rising from 334 to 973 m over 5 km. Precipitation also increases from 350 to 650 mm over this distance. Soils are derived from loess deposited over basalt bedrock (Fig. 4). The terrain is characterized by a ridge and valley pattern. Prevailing wind is from the southwest such that depth to bedrock is <60 cm for Gwin soils on windward slopes and >150 cm for Palouse soils on leeward slopes. As shallow rooted crops are more vulnerable to leaching than deeper rooted crops, the nitrate leaching potential under high rainfall is higher in the shallow soils where soluble N cannot be readily taken up by the crop. Leaching is also important for environmental reasons as the farm is situated next to the Walla Walla River – a source of drinking water for Milton-Freewater.

Historically, a single, uniform-rate application to an entire field would give little yield on the shallow soils. Herb March would compensate by fertilizing with N at 34 kg ha^{-1} on shallow soil, 56 kg ha^{-1} on moderately deep soil and 78 kg ha^{-1} on deep soil. If it was a wet year, he would apply 34 kg N ha^{-1} to a whole field, which would give this much plus residual soil N on deep soil and compensate for what leached out of the shallow soil but would only yield about 34 kg of grain ha^{-1}. In 2010, he started using mid-season aerial NIR imagery and soil EC maps to better define the boundaries of different soils. Four management zones were developed to receive 40, 47, 72 and 112 kg N ha^{-1} in order of shallow, moderately shallow, moderately deep and deep soils. A custom-made, 9.7 m wide, hoe-type, no-till drill applied seed and liquid fertilizer in one pass. Fertilizer is applied as a mixture of urea ammonium nitrate and thiosulfate (28-00-00-5), and can be varied in 9.4 L ha^{-1} increments between 94 and 375 L ha^{-1}. Ammonium phosphate (10-34-00) and potassium chloride (00-00-60) in liquid form were also applied at constant rates of 19 and 28 L ha^{-1}.

Knowing that GPC is a useful post-harvest indicator of nitrogen nutrition sufficiency in wheat (Engel et al., 1999), the farmer equipped each of his three combine harvesters with an AgLeader mass flow yield monitor and DSquared ProSpectra Grain Quality Analyzer (Fig. 5). Resulting yield and protein maps provided information for monitoring the effectiveness of his site-specific N management system. Yield loss was avoided when the N application was reduced on shallow ground and yield was increased by putting more N on deeper ground that soil tests indicated needed more N. Varying the N rate by soil depth improved the uniformity of protein levels from field to field despite a wide range in precipitation across the farm (Table 1). Protein levels of around 110 g kg^{-1} informed Herb March that his system of N management had fulfilled its purpose. He also felt that the savings in fertilizer and increase in yield more than offset the added costs of the technology in the first year.

4 Conclusions and future trends

Different site-specific nutrient management systems have developed since the mid-1980s when GPS, variable-rate equipment and other enabling technologies of precision agriculture were first introduced to farmers. In general, these systems can be divided into approaches that are either map-based or sensor-based. Map-based approaches are informed by processes that occur either in the previous season or start of the current season for example grid soil sampling or zone soil sampling. For the latter, management zones are established based on information that is gathered on certain biophysical factors that determine crop yield or simply crop yield itself. Sensor-based approaches involve optically sensing plant N sufficiency and controlling fertilizer applications in the same operation during the growing season and do not necessarily require GPS guidance.

A review of empirical studies on variable-rate fertilizer application to corn in the U.S. Midwest showed either higher or lower profitability such that economic feasibility still remains an open question (Schimmelpfennig, 2016). Economic studies in the semi-arid U.S. West are limited in number with results that are either negative (Carr et al., 1991; Long et al., 2014) or positive (Fiez et al., 1994; Beckie et al., 1997). Bongiovanni and Lowenberg-Deboer (2004) showed that site-specific N management can maintain profitability even when N is restricted to less than half the recommended uniform rate. In addition, their review of 14 studies between 1994 and 2003 revealed that variable-rate N application improved N use efficiency and reduced N leaching losses in most cases. However, many of these studies only documented reduced fertilizer use, which led to the conclusion that field studies are needed that directly measure NO_3-N losses to confirm environmental impacts of precision agriculture.

Changes in precipitation induced by climate change are projected to substantially increase N loading and threaten water resources with eutrophication across the globe (Sinha et al., 2017). The impacts are expected to be especially strong in the U.S. Corn Belt, India, China and Southeast Asia. For the U.S. Corn Belt, a 33% reduction in N inputs would be required to offset the 19% N load increase attributable to increases in precipitation that have been observed to be occurring. Site-specific nutrient management promises to help manage eutrophication and associated water quality problems but it must be capable of accounting for the impact of changing precipitation patterns on nutrient loading.

The optimization of crop production and nutrient use efficiency is a complex problem that requires a complex set of solutions to achieve improvement (Hawkesford, 2014). For example, application of N is spatially and temporally maximized when used in conjunction with germplasm optimized for traits relating to N use efficiency rather than yield alone. Other examples include (1) an increase in soluble P runoff entering Lake Erie that is an unintended

Published by Burleigh Dodds Science Publishing, 2019.

consequence of no-till farming with surface applications of P fertilizer (Jarvie et al., 2017) and (2) accelerated soil acidification in the Great Plains resulting from high rates of ammonia-based fertilizers in combination with increased use of no-till (Obour et al., 2017). In these latter two examples, no-till alone may not sufficiently improve either water or soil quality without innovative cultural practices including site-specific nutrient management to avoid applying fertilizer at rates greater than expected benefits to the crop.

The traditional mass-balance approach of establishing a yield goal and applying N for expected yield with allowance for residual soil N at or near planting has proven unreliable (Raun et al., 2017). Inaccurate N recommendations for corn are the result of N losses from soil resulting from leaching and denitrification (Morris et al., 2018). In addition, the mass-balance approach fails to account for the N mineralized, which varies within the landscape with changes in soil organic carbon content (Dharmakeerthi et al., 2005). The Maximum Return to Nitrogen (MTRN) recommendation approach (available online at http://cnrc.agron.iastate.edu/, verified 24 March 2018) was developed for corn to account for N mineralization by means of updated N-response datasets derived from an extensive network of ongoing field trials (Sawyer et al., 2006). Though useful, the MTRN approach does not address year-to-year variability in temperature or rainfall (Shanahan et al., 2008). Specialized soil test procedures are available to approximate soil N supply (Bundy and Andraski, 1995; Magdoff et al., 1984; Mulvaney et al., 2001) but would be cost-prohibitive for site-specific nutrient management (Ma and Dwyer, 1999). Future N recommendation methods are needed that reliably estimate the probability of an accurate N recommendation at the field or subfield level (Morris et al., 2018).

Under semi-arid conditions, active optical sensor-based approaches are unreliable because relationships between crop spectral reflectance and crop N status are confounded by crop variation that is largely determined by plant available water. In addition, the spectral signal measured by conventional active optical sensors will be dominated by soil spectral reflectance due to limited crop cover. Combined spectral indices that isolate the vegetation signal from the integrated reflectance within the field of view of the sensor have been found to outperform single indices such as NDVI (Eitel et al., 2009) and should be incorporated into current systems. Furthermore, green scanning LiDAR (light detection and ranging) instruments emit narrowband point illumination and promise to alleviate errors as green (532 nm) point return intensity data can be used to directly quantify crop volume and derive N concentration from single leaves (Eitel et al., 2011, 2014). Mobile, multi-wavelength LiDAR sensors may be well-suited for dryland environments and represent a possible new generation of instruments for active optical crop sensing.

Yield monitors on combine harvesters make it possible to quantify spatial variability in yield across fields but do not reveal the sources of the variability.

Since it is now possible to map grain protein levels in fields at the same spatial resolution as yield, studying the local relationship between yield and protein offers a way to better interpret yield maps. Furthermore, grain yield and protein maps can reveal the location and extent of areas within fields where more (or less) N was needed for yield more effectively than soil N maps derived from a limited sample number of soil test data. Adoption of this new technology will require industry support and scientific explanation. Commercial service providers will be needed to calibrate and maintain the on-combine spectroscopic instruments. Growers and their advisers will need specialized software that will process and remove errors from yield and protein data, and turn on-combine readings into task control maps for variable-rate application.

Process-oriented models offer great promise for prediction of plant available N and plant N demand under varying weather conditions. Van Alphen (2002) used a dynamic simulation model and real-time weather data to monitor soil N mineralization across a field and time the application of N fertilizer to wheat. Melkonian et al. (2008) developed the cloud-based computer simulation model, Adapt-N, that uses soil information and weather records to predict crop N uptake and optimum N application rates over time. In 2017, developers of Adapt-N were awarded the $1M Tulane University Nitrogen Reduction Challenge as being the most feasible and effective solution to reducing N and increasing use efficiency (http://www2.tulane.edu/tulaneprize/waterprize/index.cfm). It has been licensed to Yara International ASA - a large fertilizer company that manufactures a tractor-mounted, optical on-the-go system for in-season N management.

Though process-based models provide for more dynamic and locally adaptive N rate recommendations (Morris et al., 2018), input data may contain inaccuracies that can substantially affect N fertilizer predictions (Sela et al., 2017). Furthermore, when commercialized, these tools will cost farmers but will be lower than variable-rate methods that involve soil/tissue testing and sensor-based approaches (Morris et al., 2018). Similar to MTRN, Morris et al. (2018) proposed a framework for improving N fertilizer rate recommendations for corn that involves creation of a database with results from numerous field-scale strip trials evaluating N response with associated metadata about field history and N management practices. They advise that such a system must be practical and inexpensive so that farmers can use it on a routine basis.

In addition, Bayesian causal-learning approaches have been developed to deal with the large uncertainty facing environmental decision making. Utilizing a long-term database on crop N response that considers previous season's information, a spatio-temporal Bayesian updating method has achieved optimal efficiency under uncertainties of weather and markets when applied to variable-rate N application on dryland wheat in Montana (Lawrence et al., 2015). More recently, the method of Bayesian updating was applied to combining

Published by Burleigh Dodds Science Publishing, 2019.

in-season plant N sensing and prior information about the response to N for a given field (McFadden et al., 2018). These Bayesian approaches may provide a tractable way to improve risk prediction and decision support that are needed in site-specific nutrient management.

Variable-rate control maps utilize polygons to represent areas such as the boundaries of management zones yet crops and soils vary continuously within fields. Managing nutrients in crop production systems when biophysical properties and processes change incrementally at both short and long distances suggests the need for representing soil and landscape variation as a continuum (N. Kitchen, pers. comm.). Since many soil properties that impact crop growth can vary at short distances, the most ideal representation of spatial patterning is through a continuum map showing gradual changes. Tools and information management systems are needed to represent the soil this way along with the management decision tools and electronic variable-rate controllers that can accept continuous data.

Geophysical sensing, involving the use of magnetometers, electrical resistance meters, ground-penetrating radar and electromagnetic conductivity meters is capable of representing soil variability as a continuum. Due to high-density surveys of large areas, mobile soil sensing can also improve overall accuracy of the measurement process, even if point accuracy is less than laboratory analysis (Sudduth et al., 1997). Various approaches for sensor data fusion of key physical and chemical soil properties are available with commercial systems (Adamchuk et al., 2004). On-combine multi-sensor systems have been described for simultaneous measurement of grain yield, GPC and straw yield that can be combined to identify areas of environmental stress in fields (Long and McCallum, 2013). The fusion of multiple sensors promises to form an integral component of precision agriculture but efficient data analysis methods are needed to handle the sheer volume of data that is generated (Mahmood et al., 2012).

A significant yet overlooked source of variability within agricultural fields arises from past land use and soil fertilizer management. In the U.S. Midwest, farm fields over time expanded in size and increased adoption of machinery that enabled farmers to do more work. Certain farmstead areas were also vacated and put into crop production. Areas where animals were confined had higher amounts of manure applied. Today, those historical field boundaries and farmsteads are reflected in grid-sampled soil fertility data. Abrupt changes in soil fertility are aligned with boundaries of these historic fields and plant available P and K are highly variable on these fields (Lerch et al., 2005). Past cultural practices and management leave residual effects and impact soil physio-chemical properties for many years. This soil nutrient variability makes it difficult to develop nutrient management plans based on randomly selected samples that would be combined for single analysis, or even if a limited number of grid soil samples were collected and analysed separately. Specialized mapping methods are needed for finding and accommodating these anomalous areas.

5 Where to look for further information

A series of Site-Specific Management Guidelines is available online at http://www.ipni.net/ssmg (International Plant Nutrition Institute) where readers can find information on many aspects of site-specific nutrient management. The 4-element 'right product, right rate, right place and right time' strategy is thoroughly described in *4R Plant Nutrition: A manual for improving the management of plant nutrition* published by the International Plant Nutrition Institute. Informative book chapters have been published on site-specific nutrient management that were written in the past by prominent individuals in the field. These include Long and Pierce (2010), Raun and Schepers (2008), Haneklaus and Schnug (2006) and Pierce and Nowak (1999). See References for full details.

Centres of expertise include:

- Nitrogen Use Efficiency Web (http://nue.okstate.edu/).
- University of Missouri, Plant Sciences Extension – Nutrient management (https://plantsciences.missouri.edu/nutrientmanagement/).
- USDA Agricultural Research Service, Nutrient use and outcome network (NUOnet) (https://www.ars.usda.gov/anrds/nuonet/nuonet-home/).

6 References

Abendroth, L. J., Elmore, R. W., Boyer, M. J. and Marley, S. K. (2011), 'Corn growth and development', PMR 1009, Iowa State University Extension, Ames, IA.

Adamchuk, V., Hummel, J., Morgan, M. and Upadhyaya, S. (2004), 'On-the-go soil sensors for precision agriculture', *Computers and Electronics in Agriculture*, 44, 71–91.

Anderson-Cook, C., Alley, M., Roygard, J., Khosla, R., Noble, R. and Doolittle, J. (2002), 'Differentiating soil types using electromagnetic conductivity and crop yield maps', *Soil Science Society of America Journal*, 66, 1562–70.

Andresen, J., Hilberg, S. and Kunkel, K. (2012), 'Historical climate and climate trends in the Midwestern USA', U.S. National Climate Assessment Midwest Technical Report, Available from the Great Lakes Integrated Sciences and Assessments Center, Available online at: http://glisa.msu.edu/media/files/NCA/MTIT_Historical.pdf (last accessed 28 May 2018).

Baker, N. and Capel, P. (2011), 'Environmental factors that influence the location of crop agriculture in the conterminous United States', Scientific Investigations Report 2011-5108, U.S. Geological Survey.

Beckie, H. J., Moulin, A. P. and Pennock, D. J. (1997), 'Strategies for variable rate N fertilization in hummocky terrain', *Canadian Journal of Soil Science*, 77, 589–95.

Bermudez, M. and Mallarino, A. (2007), 'Impacts of variable-rate phosphorus fertilization based on dense grid soil sampling on soil-test phosphorus and grain yield of corn and soybean', *Agronomy Journal*, 99, 822–32.

Biermacher, J., Epplin, F., Brorsen, B., Solie, J. and Raun, W. (2009), 'Economic feasibility of site-specific optical sensing for managing nitrogen fertilizer for growing wheat', *Precision Agriculture*, 10, 213–30.

Blackmer, T., Schepers, J., Varvel, G. and Meyer, G. (1996), 'Analysis of aerial photography for nitrogen stress within corn fields', *Agronomy Journal*, 88, 729–33.

Blackmore, S. (2000), 'The interpretation of trends from multiple yield maps', *Computers and Electronics in Agriculture*, 26, 37–51.

Bongiovanni, R. and Lowenberg-Deboer, J. (2004), 'Precision agriculture and sustainability', *Precision Agriculture*, 5, 359–87.

Boydell, B. and McBratney, A. (2002), 'Identifying potential within-field management zones from cotton-yield estimates', *Precision Agriculture*, 3, 9–23.

Boyer, C., Brorsen, B., Solie, J. and Raun, W. (2011), 'Profitability of variable rate nitrogen application in wheat production', *Precision Agriculture*, 12, 473–87.

Bundy, L. and Andraski, T. (1995), 'Soil yield potential effects on performance of soil nitrate tests', *Journal of Production Agriculture*, 8, 561–8.

Bushong, J., Mullock, J., Miller, E., Raun, W. and Arnall, D. (2016a), 'Evaluation of mid-season sensor based nitrogen fertilizer recommendations for winter wheat using different estimates of yield potential', *Precision Agriculture*, 17, 470–87.

Bushong, J., Mullock, J., Miller, E., Raun, W., Klatt, A. and Arnall, D. (2016b), 'Development of an in-season estimate of yield potential utilizing optical crop sensors and soil moisture data for wheat', *Precision Agriculture*, 17, 451–69.

Cao, Q., Miao, Y., Feng, G., Gao, X., Liu, B., Li, F., Khosla, R., Mulla, D. J. and Zhang, F. (2017), 'Improving nitrogen use efficiency with minimal environmental risks using an active canopy sensor in a wheat-maize cropping system', *Field Crops Research*, 214, 365–72.

Carr, P., Carlson, G., Jacobsen, J., Nielsen, G. and Scogley, E. (1991), 'Farming soils, not fields: A strategy for increasing fertilizer profitability', *Journal of Production Agriculture*, 4, 57–61.

Castle, E., Becker, M. and Smith, F. (1972), *Farm Business Management*, Collier-Macmillan, London, UK.

Chang, J., Clay, D., Carlson, C., Clay, S., Malo, D., Berg, R., Kleinjan, J. and Wiebold, W. (2003), 'Different techniques to identify management zones impact nitrogen and phosphorus sampling variability', *Agronomy Journal*, 95, 1550–9.

Chang, J., Clay, D., Carlson, C., Reese, C., Clay, S. and Ellsbury, M. (2004), 'Defining yield goals and management zones to minimize yield and nitrogen and phosphorus fertilizer recommendation errors', *Agronomy Journal*, 96, 825–31.

Corwin, D. and Lesch, S. (2003), 'Application of soil electrical conductivity to precision agriculture', *Agronomy Journal*, 95, 455–71.

Curran, P. (1989), 'Remote sensing of foliar chemistry', *Remote Sensing of Environment*, 30, 271–8.

Delgado, J. and Follett, R. (2010), *Advances in Nitrogen Management for Water Quality*, Soil Water Conservation Society of America, Ankeny, IA.

Deming, W. (1950), *Elementary Principles of the Statistical Control of Quality: A Series of Lectures*, Nippon Kagaku Gijutsu Remmei, Tokyo, Japan.

Dharmakeerthi, R., Kay, B. and Beauchamp, E. (2005), 'Factors contributing to changes in plant available nitrogen across a variable landscape', *Soil Science Society of America Journal*, 69, 453–62.

Dhital, S. and Raun, W. (2016), 'Variability in optimum nitrogen rates for maize', *Agronomy Journal*, 108, 2165–73.

Doerge, T. (2000), 'Management zone concepts. Site-specific management guidelines', International Plant Nutrition Institute, Available online at: http://www.ipni.net/publication/ssmg.nsf/0/C0D052F04A53E0BF852579E500761AE3/$FILE/SSMG-02.pdf (last accessed 27 March 2018).

Eitel, J., Long, D., Gessler, P., Hunt, E. and Brown, D. (2009), 'Sensitivity of ground-based remote sensing estimates of wheat chlorophyll content to variation in soil reflectance', *Soil Science Society of America Journal*, 73, 1715–23.

Eitel, J., Vierling, L., Long, D. and Hunt, E. (2011), 'Early season remote sensing of wheat nitrogen status using a green scanning laser', *Agricultural and Forest Meteorology*, 151, 1338–45.

Eitel, J., Magney, T., Vierling, L., Brown, T. and Huggins, D. (2014), 'LiDAR based biomass and crop nitrogen estimates for rapid, non-destructive assessment of wheat nitrogen status', *Field Crops Research*, 159, 21–32.

Engel, R., Long, D., Carlson, G. and Meier, C. (1999), 'Method for precision nitrogen management in spring wheat: I. Fundamental relationships', *Precision Agriculture*, 1, 327–38.

Farid, H., Bakhsh, A., Ahmad, N., Ahmad, A. and Mahmood-Khan, Z. (2015), 'Delineating site-specific management zones for precision agriculture', *Journal of Agricultural Science*, 154, 273–86.

Fiez, T., Miller, B. and Pan, W. (1994), 'Assessment of spatially variable nitrogen fertilizer management in winter wheat', *Journal of Production Agriculture*, 7, 86–93.

Fleming, K., Heerman, D. and Westfall, D. (2004), 'Evaluating soil color with farmer input and apparent soil electrical conductivity for management zone delineation', *Agronomy Journal*, 96, 1581–7.

Flowers, M., Weisz, R. and White, J. (2005), 'Yield-based management zones and grid sampling strategies: Describing soil test and nutrient variability', *Agronomy Journal*, 97, 968–82.

Frankson, R., Kunkel, K., Champion, S. and Stewart, B. (2017), 'Missouri state summary', NOAA Technical Report NESDIS 149-MO, Available online at: https://statesummaries.ncics.org/mo (last accessed 27 March 2018).

Franzen, D. W., Cihacek, L. J., Hofman, V. L. and Swenson, L. J. (1998), 'Topography-based sampling compared with grid sampling in the Northern Great Plains', *Journal of Production Agriculture*, 11, 364–70.

Franzen, D., Hopkins, D., Sweeney, M., Ulmer, M. and Halvorson, A. (2002), 'Evaluation of soil survey scale for zone development of site-specific nitrogen management', *Agronomy Journal*, 94, 381–9.

Fridgen, J.J. (2000), 'Development and evaluation of unsupervised clustering software for sub-field delineation of agricultural fields', M.S. thesis, University of Missouri, Columbia, MO.

Fridgen, J., Kitchen, N., Sudduth, K., Drummond, S., Wiebold, W. and Fraisse, C. (2004), 'Management zone analyst (MZA): Software for subfield management zone delineation', *Agronomy Journal*, 96, 100–8.

Published by Burleigh Dodds Science Publishing, 2019.

Glenn, D. M., Carey, A., Bolton, F. E. and Vavra, M. (1985), 'Effect of N fertilizer on protein content of grain, straw, and chaff tissues in soft white winter wheat', *Agronomy Journal*, 77, 229–32.

Goos, R. J., Westfall, D. G., Ludwick, A. E. and Goris, J. E. (1982), 'Grain protein as an indicator of N sufficiency for winter wheat', *Agronomy Journal*, 74, 130–3.

Haneklaus, S. and Schnug, E. (2006), 'Site-specific nutrient management: Objectives, current status, and future research needs', in Srinivasan, A. (Ed.), *Handbook of Precision Agriculture: Principles and Applications*, The Haworth Press, Inc., Binghamton, NY.

Hawkesford, M. (2014), 'Reducing the reliance on nitrogen fertilizer for wheat production', *Journal of Cereal Science*, 59, 276–83.

Holland, K. and Schepers, J. (2010), 'Derivation of a variable rate nitrogen application model for in-season fertilization of corn', *Agronomy Journal*, 102, 1415–24.

Horler, D., Dockray, M. and Barber, J. (1982), 'The red edge of plant leaf reflectance', *International Journal of Remote Sensing*, 4, 273–88.

Hornung, A., Khosla, R., Reich, R., Inman, D. and Westfall, D. (2006), 'Comparison of site-specific management zones: Soil-color-based and yield-based', *Agronomy Journal*, 98, 407–15.

Jarvie, H., Johnson, L., Sharpley, A., Smith, D., Baker, D., Bruulsema, T. and Confesor, R. (2017), 'Increased soluble phosphorus loads to Lake Erie: Unintended consequences of conservation practices?' *Journal of Environmental Quality*, 46, 123–32.

Johnson, C., Mortensen, D., Wienhold, B., Shanahan, J. and Doran, J. (2003), 'Site-specific management zones based on soil electrical conductivity in a semiarid cropping system', *Agronomy Journal*, 95, 303–15.

Johnston, A. and Bruulsema, T. (2014), '4R nutrient stewardship for improved nutrient use efficiency', *Procedia Engineering*, 83, 365–70.

King, J., Dampney, P., Lark, R., Wheeler, H., Bradley, R. and Mayr, T. (2005), 'Mapping potential crop management zones within fields: Use of yield-map series and patterns of soil physical properties identified by electromagnetic induction sensing', *Precision Agriculture*, 6, 167–81.

Kitchen, N., Sudduth, K. and Drummond, S. (1999), 'Soil electrical conductivity as a crop productivity measure for claypan soils', *Journal of Production Agriculture*, 12, 607–17.

Kitchen, N., Drummond, S., Sudduth, K. and Buchleiter, G. (2003), 'Soil electrical conductivity and topography related to yield for three contrasting soil-crop systems', *Agronomy Journal*, 95, 483–95.

Kitchen, N., Sudduth, K., Myers, D., Drummond, S. and Hong, S. (2005), 'Delineating productivity zones on claypan soils using apparent soil electrical conductivity', *Computers and Electronics in Agriculture*, 46, 285–308.

Kitchen, N., Sudduth, K., Drummond, S. and Vories, E. (2010), 'Ground-based canopy reflectance sensing for variable-rate nitrogen corn fertilization', *Agronomy Journal*, 102, 71–84.

Kozar, B., Lawrence, R. and Long, D. (2002), 'Soil phosphorus and potassium mapping using a spatial correlation model incorporating terrain slope gradient', *Precision Agriculture*, 3, 407–17.

Kramer, A. Doane, T., Horwath, W. and van Kessel, C. (2002), 'Short-term nitrogen-15 recovery vs. long-term total soil N gains in conventional and alternative cropping systems', *Soil Biology and Biochemistry*, 34, 43–50.

Lawrence, P., Rew, L. and Maxwell, B. (2015), 'A probabilistic Bayesian framework for progressively updating site-specific recommendations', *Precision Agriculture*, 16, 275–96.

Lerch, R., Kitchen, N., Kremer, R., Donald, W., Alberts, E., Sadler, J., Sudduth, K. A., Myers, D. B. and Ghidey, F. (2005), 'Development of a conservation-oriented precision agriculture system: Water and soil quality assessment', *Journal of Soil and Water Conservation*, 60, 411–21.

Liggins, M., Hall, D. and Llinas, J. (2001), *Handbook of Multisensor Data Fusion*, CRC Press, Boca Raton, FL.

Long, D. and McCallum, J. (2013), 'Mapping straw yield using on-combine light detection and ranging (lidar)', *International Journal of Remote Sensing*, 34, 6121–34.

Long, D. and Pierce, F. (2010), 'Precision farming for nitrogen management', in J. A. Delgado and R. F. Follett (Eds), *Advances in Nitrogen Management for Water Quality*, Soil Water Conservation Society of America, Ankeny, IA.

Long, D., Carlson, G. and DeGloria, S. (1995), 'Quality of field management maps', in P. C. Robert, R. H. Rust and W. E. Larson (Eds), *Proceedings of 2nd International Conference on Site-specific Management for Agricultural Systems*, ASA-CSSA-SSSA, Madison, WI, pp. 251–72.

Long, D., Engel, R. and Carlson, G. (2000) 'Method for precision nitrogen management in spring wheat: II. Implementation', *Precision Agriculture*, 2, 25–38.

Long, D., Engel, R. and Carpenter, F. (2005) 'On-combine sensing and mapping of wheat protein concentration', *Crop Management*, 4(1), Available online at https://pubag.nal.usda.gov/pubag/downloadPDF.xhtml?id=11894&content=PDF (last accessed 28 May 2018).

Long, D., Engel, R. and Siemens, M. (2008), 'Measuring grain protein concentration with in-line near infrared reflectance spectroscopy', *Agronomy Journal*, 100, 247–52.

Long, D., Whitmus, J., Engel, R. and Brester, G. (2014), 'Net returns from terrain-based variable rate nitrogen management on dryland spring wheat in northern Montana', *Agronomy Journal*, 107, 1055–67.

Long, D., McCallum, J. D., Reardon, C. L. and Engel, R. E. (2017), 'Nitrogen requirement to change protein concentration of spring wheat in Pacific Northwest', *Agronomy Journal*, 109, 675–83.

Lory, J. A. and Scharf, P. C. (2003), 'Yield goal versus delta yield for predicting fertilizer nitrogen need in corn', *Agronomy Journal*, 95, 994–9.

Lowrance, C., Fountas, S., Laikos, V. and Vellidis, G. (2016), 'EZZone – An online tool for delineating management zones', *Proceedings of 13th International Conference on Precision Agriculture*, Available online at: http://vellidis.org/wp-content/uploads/2016/07/Lowrance-Paper1931-ICPA-2016.pdf (last accessed 28 May 2018).

Ma, B. and Dwyer, L. (1999), 'Within plot variability in available soil mineral N in relation to leaf greenness and yield', *Communications in Soil Science and Plant Analysis*, 30, 1919–28.

Magdoff, F., Ross, D. and Amadon, J. (1984), 'A soil test for nitrogen availability to corn', *Soil of America Journal*, 48, 1301–4.

Mahmood, H., Hoogmoed, W. and van Henten, E. (2012), 'Sensor data fusion to predict multiple soil properties', *Precision Agriculture*, 13, 628–45.

Published by Burleigh Dodds Science Publishing, 2019.

Mallarino, A. and Wittry, D. (2001), 'Management zones soil sampling: A better alternative to grid and soil type sampling?' in *Proceedings of 13th Annual Integrated Crop Management Conference*, Iowa State University Extension, Ames, IA, pp. 159–64, Available online at http://www.agronext.iastate.edu/soilfertility/info/ICM_2001_ZoneSampling_Publ.pdf (last accessed 28 May 2018).

Mausbach, M. and Wilding, L. (1991), *Spatial Variability of Soils and Landforms*, Soil Science Society of America Spec. Pub. 28, Soil Science Society of America, Madison, WI.

McCann, B. L., Pennick, D. J., van Kessel, C. and Walley, F. L. (1996), 'The development of management units for site-specific farming', in P. C. Robert, R. H. Rust and W. E. Larson (Eds), *Proceedings of 3rd International Conference on Precision Agriculture*, ASA-CSSA-SSSA, Madison, WI, pp. 295–302.

McFadden, B., Brorsen, B. and Raun, W. (2018), 'Nitrogen fertilizer recommendations based on plant sensing and Bayesian updating', *Precision Agriculture*, 19, 79-92.

Melkonian, J., van Es, H., DeGaetano, A. and Joseph, L. (2008), 'ADAPT-N: Adaptive nitrogen management for maize using high resolution climate data and model simulations', in R. Khosla (Ed.), *Proceedings of the 9th International Conference on Precision Agriculture*, International Society of Precision Agriculture, Monticello, IL, Available online at: https://cpb-us-e1.wpmucdn.com/blogs.cornell.edu/dist/8/6785/files/2016/06/Prec-Ag-Conf-2008-Melkonian-van-Es-uhaslu.pdf (last accessed 28 May 2018).

Milfred, C. and Kiefer, R. (1976), 'Analysis of soil variability with repetitive aerial photography', *Soil Science Society of America Journal*, 40, 553–7.

Milne, A., Webster, R., Ginsburg, D. and Kindred, D. (2012), 'Spatial multivariate classification of an arable field into compact management zones based on past crop yields', *Computers and Electronics in Agriculture*, 80, 17–30.

Morris, T., Murrell, T., Beegle, D., Camberato, J., Ferguson, R., Grove, J., Ketterings, Q., Kyveryga, P. M., Laboski, C. A. M., McGrath, J. M., Meisinger, J. J., Melkonian, J., Moebius-Clune, B. N., Nafziger, E. D., Osmond, D., Sawyer, J. E., Scharf, P. C., Smith, W., Spargo, J. T., van Es, H. M. and Yang, H. (2018), 'Strengths and limitations of nitrogen rate recommendations for corn and opportunities for improvement', *Agronomy Journal*, 110, 1–37.

Moulin, A. P., Anderson, D. W. and Mellinger, M. (1994), 'Spatial variability of wheat yield, soil properties and erosion in hummocky terrain', *Canadian Journal of Soil Science*, 74, 219–28.

Mulvaney, R., Khan, S., Hoeft, R. and Brown, H. (2001), 'A soil organic nitrogen fraction that reduces the need for nitrogen fertilization', *Soil Science Society of America Journal*, 65, 1164–72.

Nielsen, G., Long, D. and Queen, L. (1996), 'Mapping potential of digitized aerial photographs and space images for site-specific crop management', *Proceedings of the SPIE 2818, Multispectral Imaging for Terrestrial Applications*, http://dx.doi.org/10.1117/12.256078

Nix, J. (1979), 'Farm management: The state of the art (or science)', *Journal of Agricultural Economics*, 39, 277–91.

Nolan, S. C., Goddard, T. W., Heaney, D. J., Penne, D. C. and McKenzie, R. C. (1995), 'Effects of fertilizer on yield at different soil landscape positions', in P. C. Robert, R. H. Rust and W. E. Larson (Eds), *Proceedings of 2nd International Conference on Precision Agriculture*, ASA-CSSA-SSSA, Madison, WI, pp. 553–8.

Obour, A., Mikha, M., Holman, J. and Stahlman, P. (2017), 'Changes in soil surface chemistry after fifty years of tillage and nitrogen fertilization', *Geoderma*, 308, 46–53.

Pennock, D. and Corre, M. (2001), 'Development and application of landform segmentation procedures', *Soil Tillage Research*, 69, 15–26.

Pierce, F. J. and Nowak, P. (1999), 'Aspects of precision agriculture', in D. L. Sparkes (Ed.), *Advances in Agronomy*, volume 67, Academic Press, San Diego, CA, pp. 1–85.

Ping, J. and Dobermann, A. (2005), 'Processing of yield map data', *Precision Agriculture*, 6, 193–212.

Raun, W. and Schepers, J. (2008), 'Nitrogen management for improved use efficiency', in J. S. Schepers and W. R. Raun (Eds), *Nitrogen in Agricultural Systems,* Agronomy Monograph 49, ASA-CSSA-SSSA, Madison, WI.

Raun, W., Solie, J., Johnson, G., Stone, M., Mullen, R., Freeman, K., Thomason, W. E. and Lukina, E. V. (2002), 'Improving nitrogen use efficiency in cereal grain production with optical sensing and variable rate application', *Agronomy Journal*, 94, 815–20.

Raun, W., Figueiredo, B., Dhillon, J., Fornath, A., Bushong, J., Zhang, H. and Taylor, R. (2017), 'Can yield goals be predicted?' *Agronomy Journal*, 109, 2389–95.

Rouse, J., Haas, R., Schell, J. and Deering, D. (1974), 'Monitoring vegetation systems in the Great Plains with ERTS', *in* S. C. Freden, E. P. Mercanti and M. Becker (Eds), *Third Earth Resources Technology Satellite-1 Symposium. Volume I: Technical Presentations*, NASA SP-351, NASA, Washington DC, pp. 309–17.

Sawyer, J., Nafziger, E., Randall, G., Bundy, L., Rehm, G. and Joern, B. (2006), 'Concepts and rationale for regional nitrogen rate guidelines for corn', PM 2015, Iowa State University Extension, Ames, IA.

Scharf, P. (2015), *Managing Nitrogen in Crop Production*, ASA-CSSA-SSSA, Madison, WI.

Scharf, P. and Lory, J. (2009), 'Calibrating reflectance measurements to predict optimal sidedress nitrogen rate for corn', *Agronomy Journal*, 101, 615–25.

Scharf, P., Brouder, S. and Hoeft. R. (2006), 'Chlorophyll meter readings can predict nitrogen need and yield response of corn in the north-central U.S.', *Agronomy Journal*, 98, 655–65.

Scharf, P. C., Shannon, D. K., Palm, H. L., Sudduth, K. A., Drummond, S. T., Kitchen, N. R., Mueller, L. J., Hubbard, V. C. and Oliveira, L. F. (2011), 'Sensor-based nitrogen applications out-performed producer-chosen rates for corn in on-farm demonstrations', *Agronomy Journal*, 103, 1683–91.

Schepers, J., Francis, D., Vigil, M. and Below, F. (1992), 'Comparison of corn leaf nitrogen concentration and chlorophyll meter readings', *Communications in Soil Science and Plant Analysis*, 23, 2173–87.

Schepers, A., Shanahan, J., Liebig, M., Schepers, J., Johnson, S. and Luchiari Jr., A. (2004), 'Appropriateness of management zones for characterizing spatial variability of soil properties and irrigated corn yields across years', *Agronomy Journal*, 96, 195–203.

Schimmelpfennig, D. (2016), 'Farm profits and adoption of precision agriculture', ERR-217, U.S. Department of Agriculture, Economic Research Service.

Sela, S., van Es, H., Moebius-Clune, B., Marjerison, R., Melkonian, J. and Moebius-Clune, D. (2017), 'Adapt-N recommendations for maize nitrogen management increase grower profits and reduce environmental losses on Northeast and Midwest USA farms', *Agronomy Journal*, 108, 1726–34.

Selles, F. and Zentner, R. P. (2001), 'Grain protein as a post-harvest index of N sufficiency for hard red spring wheat in the semiarid prairies', *Canadian Journal of Plant Science*, 81, 631–6.

Sembiring, H., Raun, W. R., Johnson, G. V., Stone, M. L., Solie, J. B. and Phillips. S. B. (1998), 'Detection of nitrogen and phosphorus nutrient status in winter wheat using spectral radiance', *Journal of Plant Nutrition*, 21, 1207–33.

Shanahan, J., Kitchen, N., Raun, W. and Schepers, J. (2008), 'Responsive in-season nitrogen management for cereals', *Computers and Electronics in Agriculture*, 61, 51–62.

Shewhart, W. (1939), *Statistical Method from the Viewpoint of Quality Control*, Graduate School of the Department of Agriculture, Washington DC.

Sinha, E., Michalak, A. and Balaji, V. (2017), 'Eutrophication will increase during the 21st century as a result of precipitation changes', *Science*, 357, 405–8.

Solie, J. B., Raun, W. R., Whitney, R. W., Stone, M. L. and Ringer, J. D. (1996), 'Optical sensor based field element size and sensing strategy for nitrogen application', *Transactions of the American Society of Agricultural Engineers*, 39, 1986–92.

Solie, J., Monroe, A., Raun, W. and Stone, M. (2012), 'Generalized algorithm for variable-rate nitrogen application in cereal grains', *Agronomy Journal*, 104, 378–87.

Stafford, J. V., Lark, R. M. and Bolam, H. C. (1999), 'Using yield maps to regionalize fields into management units', *in* P. C. Robert, R. H. Rust and W. E. Larson (Eds), *Proceedings of the 4th International Conference on Precision Agriculture*, ASA, CSSA and SSSA, Madison, WI, pp. 225–37.

Stone, M. L., Solie, J. B., Raun, W. R., Whitney, R. W., Taylor, S. L. and Ringer, J. D. (1996), 'Use of spectral reflectance for correcting in-season fertilizer nitrogen deficiencies in winter wheat', *Transactions of the American Society of Agricultural Engineers*, 39, 1623–31.

Sudduth, K., Hummel, J., Birrell, S., (1997), 'Sensors for site-specific management', in F. J. Pierce and E. J. Sadler (Eds), *The State of Site-Specific Management for Agriculture*, ASA, CSSA and SSSA, Madison, WI.

Taylor, R., Kluitenberg, G., Schrock, M., Zhang, N., Schmidt, J. and Havlin, J. (2001), 'Using yield monitor data to determine spatial crop production potential', *Transactions of the American Society of Agricultural Engineers*, 44, 1409–14.

Van Alphen, B. (2002), 'A case study on precision nitrogen management in Dutch arable farming', *Nutrient Cycling in Agroecosystems*, 62, 151–61.

van Es, H., Kay, B., Melkonian, J. and Sogbedji, J. (2007), Nitrogen management for maize in humid regions: Case for a dynamic modeling approach, in T. Bruulsema (Ed.), *Managing Crop Nutrition for Weather*, International Plant Nutrition Institute Publication, Norcross, GA, pp. 6–13.

Von Rosenberg Jr., C. W., Abbate, A. and Drake, J. (2000), 'A rugged near-infrared spectrometer for real-time measurement of grains during harvest', *Spectroscopy*, 15, 34–8.

Wade, J., Horwath, W. and Burger, M. (2016), 'Integrating soil biological and chemical indices to predict net nitrogen mineralization across California agricultural systems', *Soil Science Society of America Journal*, 80, 1675–87.

Wollenhaupt, N., Wolkowski, R. and Clayton, M. (1994), 'Mapping soil test phosphorus and potassium for variable-rate fertilizer application', *Journal of Production Agriculture*, 7, 441–8.

Zhang, X., Shi, L., Jia, X., Seielstad, G. and Helgason, C. (2010), 'Zone mapping application for precision-farming: A decision support tool for variable rate application', *Precision Agriculture*, 11, 103–14.

Chapter 3

Developments in the use of fertilizers

Bryan G. Hopkins, Brigham Young University, USA

1 Introduction

Ancient Greek philosophers, such as Parmenides, were among the first to observe that 'ex nihilo nihil fit' or 'nothing comes from nothing'. This concept was later developed into the foundational law of the conservation of mass, stating chemical elements (mass) are neither created nor destroyed. This concept is helpful in understanding nutrient cycles. In terrestrial ecosystems, plants take up nutrients from the soil and, as they die, residues are recycled back to provide nutrients for succeeding plants (Marschner and Rengel, 2007). Nutrients can also be solubilized from sources such as rock. They can also be deposited back into the system through processes such as flooding and wind deposition. Nutrients can also be lost in these cycles through processes such as erosion, leaching, and/or volatilization and other gaseous emissions. Fertile soils are able to store nutrients to resupply plants, but this is a finite resource. If nutrient output exceeds inputs, the resupply rate, based on the principles of equilibrium chemistry and microbial degradation of residues, slows as the nutrient soil storage decreases. These natural processes of

http://dx.doi.org/10.19103/AS.2019.0062.26

nutrient cycling are generally adequate to sustain plant growth in natural biomes such as forests (assuming factors such as adequate moisture and the right temperature).

However, agricultural production imposes new pressures on the process of nutrient cycling since nutrients are also lost from the system from harvesting crops or grazing animals. The move to more intensive farming characterized by monocultures of high-yield crops has intensified these pressures, given the high nutrient requirements needed to sustain these production systems over time (Hopkins and Hansen, 2019). These more intensive systems disrupt natural nutrient cycling systems, which can then struggle to meet enhanced crop nutrient demands.

One response has been the development of fertilizers to fill the nutrient gap. A fertilizer is any 'organic or inorganic material of natural or synthetic origin (other than liming materials) that is added to a soil to supply one or more plant nutrients essential to the growth of plants' (Anon., 2008). This chapter limits this definition to mineral nutrients. Non-mineral nutrients (carbon, hydrogen, and oxygen) are often components of fertilizer materials, but plants obtain the majority of their needs for these from the air and water in the environment.

Table 1 lists some of the most important historical developments in fertilizers. The need for crop nutrition began with the first agricultural revolution (Neolithic period–circa 10 000 BCE) when humans began to settle and grow crops. As varieties and yields started to improve, farmers became aware of the need to support cultivation through both irrigation and the first use of fertilizers. The Ancient Egyptians were aware of the importance of the nutrients that came with silt deposits from the annual flooding of the Nile River. Texts from Persian, Jewish, Roman, Inca, Indus, and other ancient civilizations refer to the application of ash and animal manure to support crop growth. However, farmers were largely dependent on local-sourced materials for crop nutrition. This was a factor in yields remaining low in the ancient world. As an example, wheat (*Triticum* spp.) yields were ~0.5–1.2 Mg ha^{-1} in some of the earliest, rain-fed agricultural communities in the Near East 3000–12 000 years BCE (Araus et al., 2014; see Fig. 1). For centuries, yield increases remained modest. For example, yields in the USA prior to 1940 were approximately the same as in ancient times.

The foundations for the modern fertilizer industry go back to scientific advances from the Renaissance onward, which laid the foundations of modern chemistry. This led to the discovery of key elements involved in plant nutrition, such as phosphorus (discovered in 1669), followed by nitrogen in 1772 and potassium in 1807 (Table 2). The first systematic experiments in crop nutrition began in the 18th century focusing on crop rotations–an essential consideration in plant nutrition (Myers et al., 2008; Hopkins, 2015). In the 19th century, work by scientists at the Rothamsted Experimental Station in the UK demonstrated

Table 1 Fertilizer development–select historical events

Stone age	
Pre-history	Hunter gatherers
First agricultural revolution (Neolithic)	
10 000 BCE	Ash and manure fertilizers
500 CE	Potash use reported
1300	Potash mining in Ethiopia
Second agricultural revolution (industrial and enlightenment)	
1669	Beginning of advanced fertilizer chemistry science (phosphorus element discovered)
1736	Large-scale production of sulfuric acid
1754	Ammonia gas discovered
1770	Discovery of phosphoric acid
1772	Nitrogen element discovered
1784	Annals of Agriculture published–fertilizer experimentation
1790	First US patent is for potash refinement
1799	Discovery that N and P taken up by roots
1802	Guano a global fertilizer source
1807	Discovery of potassium element; chemically manufactured fertilizer production
1815	First synthetically produced fertilizer–ammonium sulfate
1828	First sodium nitrate manufacturing plant
1828	Urea synthesized
1830	Widespread use of mined sodium nitrate
1840	Acidification of bones and rock–birth of phosphate fertilizer industry
1843	Rothamsted Experiment Station–formal fertilizer field science
1854	First phosphate manufacturing plant
1856	Development of potash fertilizer mines in Germany
1894	First marketable elemental S produced
1898	Calcium cyanimide synthesis
1905	Nitric acid synthesis
1910	Haber-Bosch process for generating N fertilizer from N_2 gas
1913	First large-scale ammonia synthesis plant
1918	War stocks of explosive ammonium nitrate released for fertilizer use
1922	Widespread urea manufacturing
1926	Ammonium phosphate fertilizers created
1943	First field application of anhydrous ammonia
Third agricultural revolution (Green Revolution/digital information age)	
1954	First large-scale diammonium phosphate plant
1955	NPK complexes begin to be used widely

(*Continued*)

Table 1 (*Continued*)

1955	First slow-release fertilizer commercially produced (urea formaldehyde)
1961	Slow-release methylene urea produced
1967	Production of control-release polymer-coated fertilizers
1969	Environmental impacts of fertilizers gaining momentum
1974	Large-scale production of slow-release sulfur-coated fertilizers
1990	Increasing adoption of variable-rate fertilization using remote sensing
1995	Shift to more widespread use of urea due to terrorist bombing using ammonium nitrate fertilizer
2017	Widespread research and government involvement with biostimulants

BCE = before current era; CE = current era.

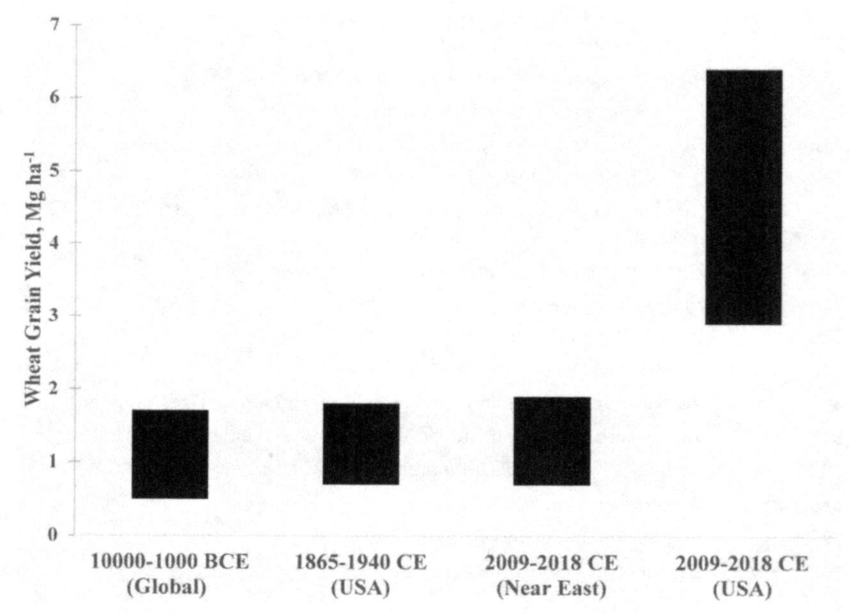

Figure 1 Range of historic wheat yields compared to those immediately prior to and after the Green Revolution (Araus et al., 2014; Hopkins and Hansen, 2019).

that yield loss due to continual cropping could be abated with relatively high rates of fertilization. At the same time, the first synthetic fertilizers began to be manufactured, initially by treating phosphate with sulfuric acid. Potash (potassium chloride) and other chemical compounds, which had been mined for centuries, began to be repurposed for fertilizers. Sodium nitrate mines began to provide nitrogen fertilizers. Deposits of nutrient-rich guano were harvested and converted into fertilizers on a large scale. In addition, it was found that compounds such as sodium nitrate could be manufactured from other raw

Table 2 Essential plant nutrients and available diagnostic tests

Element	Form taken up by plants[a]	Major source	Reported concentration ranges for crop plants (dry weight)	Year discovered	Diagnostic test available	
					Soil	Plant
Non-mineral nutrients						
Carbon (C)	Carbon dioxide (CO_2)	Atmosphere	~45%	~1694	N/A	N/A
Oxygen (O)	Oxygen (O_2)	Atmosphere, water	~43%	1774	N/A	N/A
Hydrogen (H)	Water (H_2O)	Water	~6%	1776	N/A	N/A
Primary nutrients						
Nitrogen (N)	Nitrate (NO_3^-), ammonium (NH_4^+)	Organic matter, atmosphere	0.5–6.4%	1772	1	1
Phosphorous (P)	Phosphate ($H_2PO_4^-$, $HPO_4^=$)	Soil minerals, organic matter	0.1–1.3%	1669	1	1
Potassium (K)	Potassium (K^+)	Soil minerals	0.3–14%	1807	1	1
Secondary nutrients						
Calcium (Ca)	Calcium (Ca^{++})	Soil minerals	0.03–6%	1879	1	2
Magnesium (Mg)	Magnesium (Mg^{++})	Soil minerals	0.02–5%	1755	2	2
Sulfur (S)	Sulfate ($SO_4^=$)	Organic matter, precipitation	0.08–1.4%	Ancient	3	2
Micronutrients						
Boron (B)	Borate ($H_2BO_3^-$)	Organic matter	1-200 ppm	1808	1	2
Manganese (Mn)	Manganese (Mn^{++})	Soil minerals	1-2000 ppm	1774	1	2
Zinc (Zn)	Zinc (Zn^{++})	Soil minerals, organic matter	5-400 ppm	1875	1	2

(Continued)

Table 2 (*Continued*)

Element	Form taken up by plants[a]	Major source	Reported concentration ranges for crop plants (dry weight)	Year discovered	Diagnostic test available	
					Soil	Plant
Copper (Cu)	Copper (Cu^{++})	Soil minerals, organic matter	1–60 ppm	Ancient	2	2
Iron (Fe)	Iron (Fe^{++}, Fe^{+++})	Soil minerals, organic matter	2–1820 ppm	1735	3	2
Molybdenum (Mo)	Molybdate ($MoO_4^{=}$)	Soil minerals	0.01–20 ppm	1781	4	2
Chlorine (Cl)	Chloride (Cl^-)	Precipitation	0.05–3%	1774	2	2
Nickel (Ni)	Nickel (Ni^{++})	Soil minerals	0.1–10 ppm	1751	4	4

[a] Other forms of these nutrients are present in the soil. This list includes only forms used by plants.
For diagnostic tests, 1 = yes, with reasonably good correlations and interpretation data; 2 = yes, but with minimal correlation/interpretation data available, 3 = yes, but poor correlations with yield response, 4 = no.
Source: adapted from Hopkins et al. (2020).

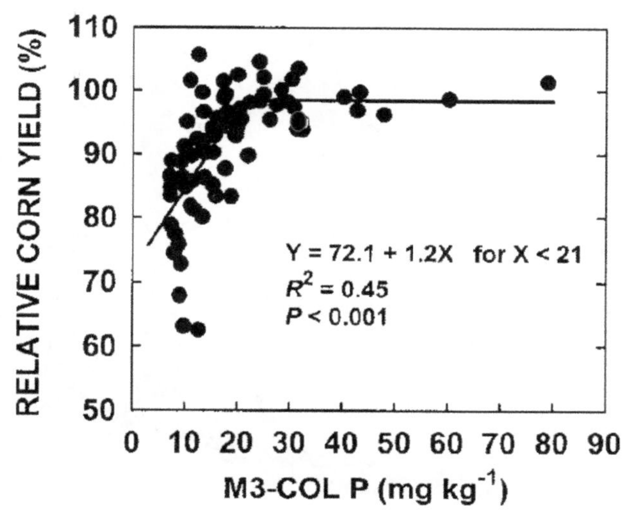

Figure 2 Example of the correlation between a proven soil extractant (Mehlich 3-colorimetric P) and yield response (Mallarino, 2003).

materials. Commercial companies selling and shipping fertilizers proliferated during the last half of the 19th and into the 20th centuries. Fertilizers began to be widely available and relatively inexpensive.

A major development in the early 20th century was the Haber-Bosch process which used natural gas to convert atmospheric nitrogen (N_2) gas into solid ammoniacal nitrogen fertilizers. This coincided with advances in soil testing with the development of extractants able to predict yield response to a majority of the nutrients applied as fertilizers (Anderson, 1960; Peck, 1990; see Fig. 2 and Table 2). Further advances included the development of mixed fertilizers such as ammonium phosphate fertilizers, which continue to be the main phosphate fertilizers used today. Other homogenous fertilizer blends, many containing all of the primary macronutrients, began to be manufactured and widely used.

These advances in fertilizers have contributed to significant improvements in yields. The Green Revolution of the 1960s was based, in part, on the development of new, high-yielding varieties and, in part, on significant increases in fertilizer use (Liu et al., 2015). The yield increases achieved in the Green Revolution are shown in Fig. 3. Record yields for each of the species shown in Fig. 3 are more than twice those of recent averages. For example, the highest yield obtained for irrigated maize in the USA in 2018 was 30 Mg ha^{-1} in comparison with average yields that were about one-third of that value (Hopkins and Hansen, 2019). This gives some indication of the potential to improve yields more broadly. Current levels of fertilizer use supporting this level of production are shown in Fig. 4.

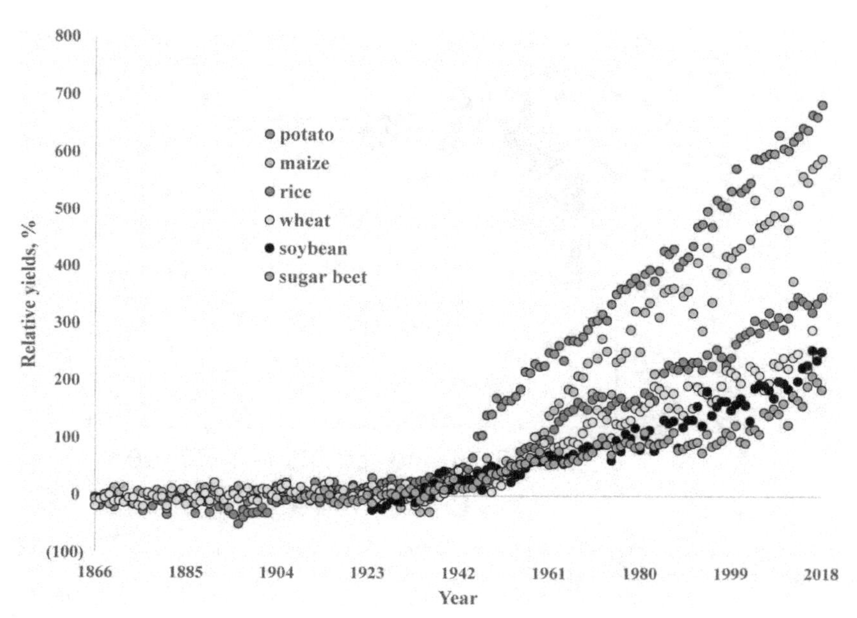

Figure 3 Annual US yields relative to the pre-Green Revolution averages (set to equal 0) of 6.4 (potato), 1.6 (maize), 1.9 (rice), 0.9 (wheat), 1.0 (soybean), and 24.8 (sugar beet) Mg ha^{-1} (Hopkins and Hansen, 2019).

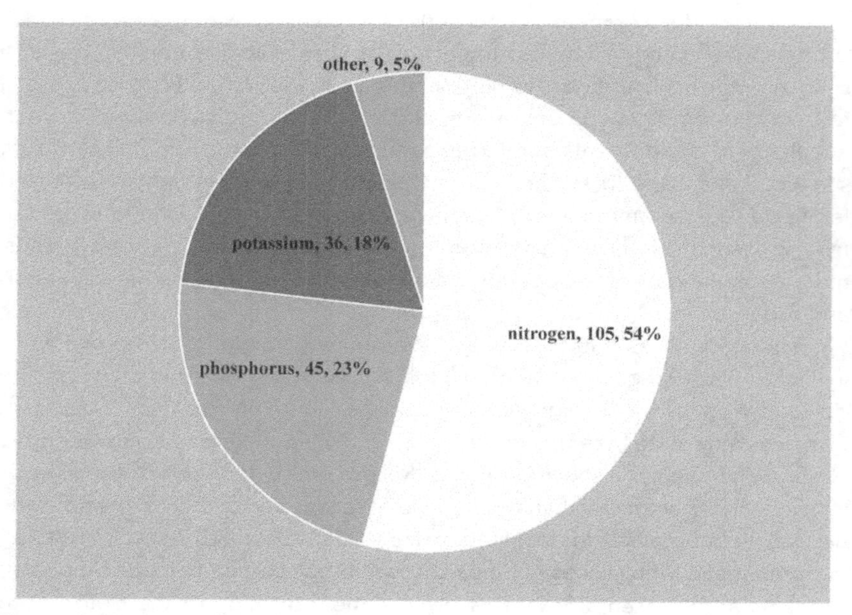

Figure 4 Nitrogen, phosphorus (as P_2O_5), potassium (as K_2O), and 'other' fertilizer consumption in 2016, as shown in million tonnes and percentage of the total. Other includes the secondary and micronutrients (Anon., 2019).

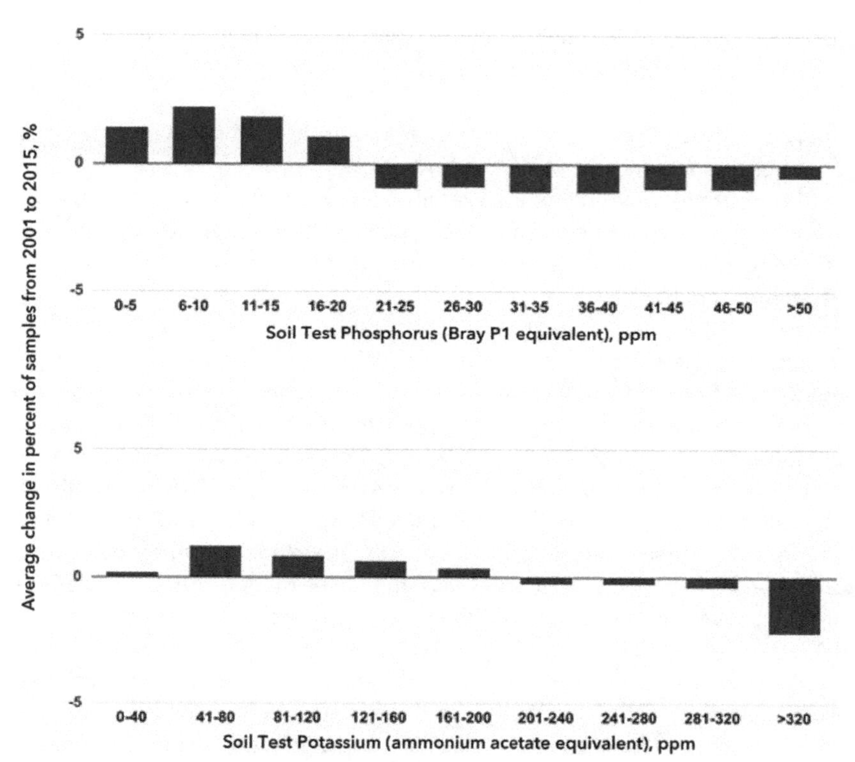

Figure 5 Change in the distribution of soil test phosphorus and potassium categories using data between 2001 and 2015 based on millions of North American soil analyses (Anon., 2015). Upward bars indicate that over the time period, there are higher percentages of soil samples within these ranges. Downward bars indicate a smaller percentage of samples within the ranges. Overall, the figure shows a depletion of soil fertility.

However, current levels of fertilization are not keeping pace with nutrient removal in many regions. Soil test levels are an indication of this trend and show that the average nutrient levels in soils in most regions in the USA are declining (with the exception of areas with high manure availability; see Fig. 5). Depleting soil fertility while simultaneously pressuring the system by continuing to increase yields is not sustainable, highlighting the importance of effective fertilization (Bruulsema et al., 2012; Hopkins and Hansen, 2019).

2 Nutrient pollution and resource depletion

The pressures toward the increasing use of fertilizers has had opportunity costs. Fertilizer use impacts the environment through resource consumption and pollution (Snyder et al., 2009; Bruulsema et al., 2012; Zhang et al., 2015;

LeMonte et al., 2016). Potential costs, discussed in the following section, include (Pierzynski et al., 2005; Zhang et al., 2015):

- Natural resource depletion, for example, from mining raw materials for fertilizer manufacture,
- Environmental pollution, for example, emissions from processes used in fertilizer manufacture or from leaching of fertilizer runoff into water bodies,
- Atmospheric emissions from fertilizers which are a contributing factor to climate change.

Nitrogen fertilizers mostly come from the Haber-Bosch process. The nitrogen itself comes from the atmosphere (which is 78% N_2 gas). Although the gas is a renewable resource, the manufacturing process is energy intensive. It has been estimated that about 1–2% of the total energy consumption on the Earth is used in this process (Skowronska and Filipek, 2014; Hasler et al., 2015). In addition to the environmental impact of their manufacture, the transport and application of fertilizers uses fossil fuels and other resources. There have been various studies to look at reducing the environmental impact of manufacture, for example, through process improvements or the use of biomass as a fuel (Ahlgren et al., 2012; Bicer et al., 2017).

Phosphorus is second only to nitrogen in its use as a fertilizer (see Fig. 4). Some claim the supply will be exhausted in a few decades, although more informed sources estimate that the supply will last several centuries as technology for recovery improves (Hopkins, 2015). Estimates of potassium reserves also suggest that they will last for centuries (Cocker et al., 2016). Nevertheless, there remains concern that future generations may run out of easily accessible reserves of raw fertilizer materials.

A more immediate problem is the negative impact of fertilizer use on the environment. Estimates of up to half of the nitrogen added to soil as fertilizer either is emitted to the atmosphere as ammonia, nitrous oxide, or other gaseous forms or finds its way into surface or ground waters as nitrate (Kibblewhite, 2007). The nitrogen cycle has gas phases that can contaminate the atmosphere (Pierzynski et al., 2005; Snyder et al., 2009; Venterea et al., 2011, 2016; Parkin and Hatfield, 2014; LeMonte et al., 2018). Nitrate in the soil can be converted to nitrous oxide, a greenhouse gas about 300 times more potent than carbon dioxide. A percentage of nitrogen fertilizer is given off as this gas. In the UK, it has been estimated that agriculture accounts for around 60% of UK emissions of nitrous oxide (Kibblewhite, 2007).

An even larger percentage of nitrogen fertilizer is lost as ammonia (commonly 5–20%), which is classified as the reactive nitrogen (Pierzynski et al., 2005; LeMonte et al., 2018). In the UK, some 90% of ammonia emissions to air are from agriculture (Kibblewhite, 2007). Nitrogen emitted as ammonia also returns to land as wet and dry deposition, sometimes causing damage to sensitive ecosystems such as heathlands or alpine areas via nutrient enrichment or as acid rain. An example of a problem with atmospheric nitrogen deposition is increasing weed species at the expense of other plants (Fernelius et al., 2017; Guthrie et al., 2018).

Nitrogen can also end up in surface water bodies and result in eutrophication. Nitrogen can accelerate algal growth, which has a damaging effect on aquatic life as microbes degrading the detritus deplete the water of dissolved oxygen. Nitrogen can also be a direct pollutant in the drinking water of mammals (Canter, 1997). It readily leaches to groundwater and, if accumulated in high enough concentrations, is potentially deadly to infants (methemoglobinemia) and is suspected of causing other health problems when found at excessively high levels.

Phosphorus is the other main fertilizer pollutant (Hopkins, 2015; Sharpley et al., 2018). Whereas nitrogen is often the limiting nutrient in coastal waters, such as in areas of the Gulf of Mexico and the Mediterranean Sea, phosphorus is typically limiting in freshwater systems. Unlike nitrogen, phosphorus is not very mobile in soil. As such, it tends to accumulate at the soil surface. Soil erosion and surface water flows over phosphorus-enriched soil can result in the accumulation of this nutrient in surface waters–effectively enriching the nutritional supply for algae and increasing its rate of growth. In the UK, about 25% of phosphate in surface waters is thought to come from agriculture (Kibblewhite, 2007), with higher estimates for the USA (Hopkins, 2015). This can result in problems with eutrophication and hypoxia, such as problems encountered in the Great Lakes in the USA and the Danube River in Europe (Aloe et al., 2014).

3 Achieving more sustainable use of fertilizers

Increasing more efficient fertilizer use can be achieved through a combination of improved plant genetics and fertilizer-use efficiency. One component is to increase nutrient-uptake efficiency in plants (Ciampitti and Vyn, 2014). As an example, modern maize hybrids are able to take up phosphorus at greater total amounts and for a more extended period during the season (Hopkins and Hansen, 2019; see Fig. 6). Research has helped to identify root distribution and other characteristics which affect nutrient-uptake efficiency and which provide potential targets for breeding improved cultivars (Gill et al., 2004; Bender et al., 2013; Andresen et al., 2016). One example is developments in potato cultivars. The most commonly grown cultivar in the USA is 'Russet Burbank', which is relatively inefficient in nutrient and water uptake due to shallow roots with poor density and few root hairs, with a phosphorus requirement that is two to three times greater than that of many other crops. This has led to the development of

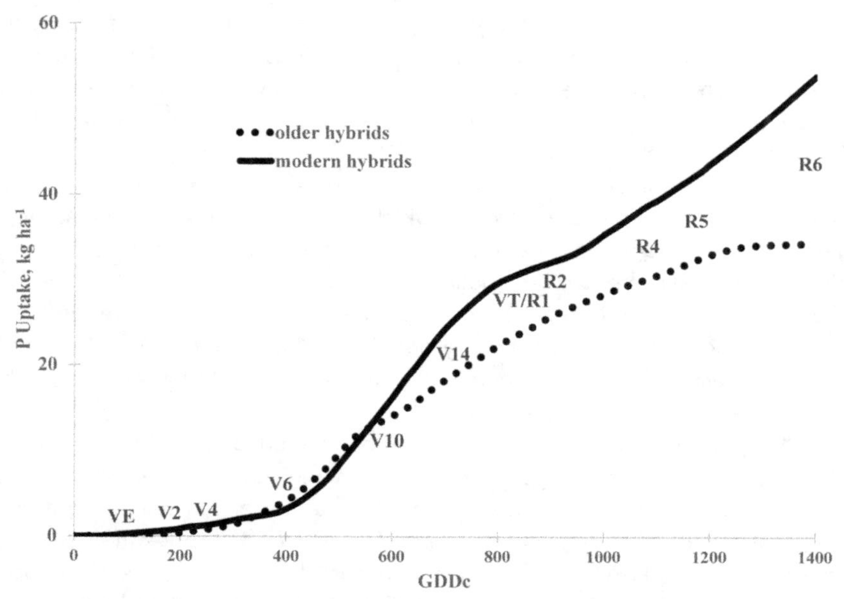

Figure 6 Estimated cumulative P uptake for maize as a function of growing degree days (GDDc) for maize growth with vegetative (V) and reproductive (R) stages shown for 'modern hybrids' and 'older hybrids' (Adapted from Bender et al., 2013; Hopkins and Hansen, 2019; Ritchie et al., 1997).

new cultivars such as 'Alturas' with greatly improved rooting efficiency (Hopkins et al., 2020).

Genetic improvements need to be coupled with more efficient and targeted use of fertilizers. Fertilizer industry is partnering with governments, farmers, and others in seeking to achieve these goals through promoting the '4R's' of fertilization: applying the Right fertilizer source at the Right rate, at the Right time, and in the Right place (Bruulsema et al., 2012). Significant research on general fertilization rates and application techniques have resulted in improvements in yields and uptake efficiency (Bruulsema et al., 2012; Hopkins and Hansen, 2019).

Other advances have been achieved through developments in precision agriculture technology. Advances include improved remote and proximal sensing technologies to identify areas of nutrient deficiency and variable-rate application (VRA) techniques to provide more targeted delivery (Stafford, 2018). Long (2018) has reviewed developments in site-specific nutrient management systems, including grid sampling, construction of management zones, in-season crop sensing, and targeted delivery systems. Variable-rate fertilization has improved both yields and efficiency in fertilizer use (Lal and Stewart, 2015). Unmanned drones and robots are the latest developments that are being developed in a wide variety of agricultural applications, including the identification of nutrient deficiencies and, potentially, highly targeted fertilizer application (e.g. direct injection into the plant; Berenstein, 2019).

Another approach has been the development of enhanced-efficiency fertilizers (EEFs) designed to improve the efficiency of nutrient uptake by timing nutrient release more closely with the plant demand (Hopkins et al., 2008). Sulfur and polymer coatings have been developed to release nutrients more slowly. Another approach has been the use of stabilizers such as urease and nitrification inhibitors, which extend the time that nutrients remain in a plant-available form (Motavalli et al., 2008). These were primarily used initially in the greenhouse industry, where the leaching of nutrients out of shallow pots was a common problem but, as costs have decreased, they have begun to be more widely used with field crops. A recent development is the use of nanoparticles to control fertilizer release (Adisa et al., 2019). A recent review has suggested that, while they are not a complete solution, EEF can play a significant role in reducing nitrogen losses to the environment (Li et al., 2018). While EEF products are more expensive, costs can be recouped by the need for less fertilizers to achieve the required yield and their application in a single operation rather than multiple applications throughout the season. Slow-release products also reduce potential toxicity to seeds or seedlings.

Fertilizers with enhanced purity and solubility have also been developed for liquid applications. The development of fertigation through sprinkler and drip-irrigation has greatly aided in improving yields and precision fertilization (Fageria et al., 2009). Foliar nutrient applications are increasing with advances in the accuracy of low-volume sprays. Adjuvant chemicals have been developed to aid foliar fertilizers being taken up more efficiently. The role of nutrients in pest management is also vital, such as interactions with pathogens (Dordas, 2008; Bensen et al., 2009).

Another major segment of fertilizer development is related to organic farming (Hopkins and Hirnyck, 2007). This movement has been around since the dawn of commercial fertilizers, with the term 'organic farming' coined in 1940. The general idea is that only naturally occurring products are used as fertilizers, such as using a previous crop of legumes for nitrogen and composted manure. However, there are many exceptions, with the final determination of what is acceptable made by various government certifying agencies.

4 Developments in nitrogen fertilizers

Of all the plant nutrients, nitrogen is sold in the largest volume because it has the most impact on yield (Geary et al., 2015; Schlegel and Havlin, 2017; Zhang et al., 2015; see Fig. 4). Nitrogen is needed in relatively large quantities in plants (see Table 2). Nitrogen deficiencies are highly visible in plant tissues and, compared to other nutrients, have more immediate and dramatic effects (see Fig. 7).

Figure 7 Nitrogen-deficient quinoa with less grain yield, stunting, changes in phenotypic traits (e.g. leaf serrations), and chlorosis (Cole et al., 2020).

There is a large store of nitrogen in the soil, mostly occurring as a component of soil organic matter (SOM). However, only about 2–5% of this is mineralized to become available for crop uptake each year. Given this and the high demand for nitrogen in plant tissues, nitrogen fertilizers often need to be applied to crops in order to achieve maximum economic yield. The rate of application is dependent on a range of factors including estimated degree of mineralization of organic matter, nitrate and ammonium concentrations in surface and subsurface soil, nitrate in irrigation water, and the application of manure or other nitrogen-rich wastes. The rate may need to be increased if a previous crop residue has relatively low nitrogen and high carbon concentrations and is incorporated close to the time of planting. Root depth, architecture, and nutrient-uptake efficiency are also important factors, with deep and more densely rooted crops often able to capture leached residual nitrogen and other nutrients from a previous crop (Andresen et al., 2016; Hopkins and Hansen, 2019).

The major traditional nitrogen fertilizers and their use from the beginning of the Green Revolution in the 1960s, when fertilizers began to be used widely, to 2016 are shown in Fig. 8. One of the key developments has been the emergence of urea as the main source for nitrogen fertilizers. This is important because there is a high loss potential from the volatilization of ammonia gas during the hydrolysis conversion from urea to ammonia gas and then ammonium. This has resulted in major efforts to minimize this loss. Another major shift is away from ammonium sulfate, which has contributed to a recent increase in sulfur deficiencies in crops.

Traditional fertilizer nitrogen is highly soluble and converts rapidly to nitrate. This can lead to leaching and denitrification/nitrification losses (Van Groenigen et al., 2010; Venterea et al., 2011, 2016; Canter, 1997). As such, timing and placement are critical for improved efficiency. It is important to understand plant-uptake patterns for nitrogen and ensure that nitrogen in plant-available forms (nitrate and ammonium) is present when plants need it. As such, there has been

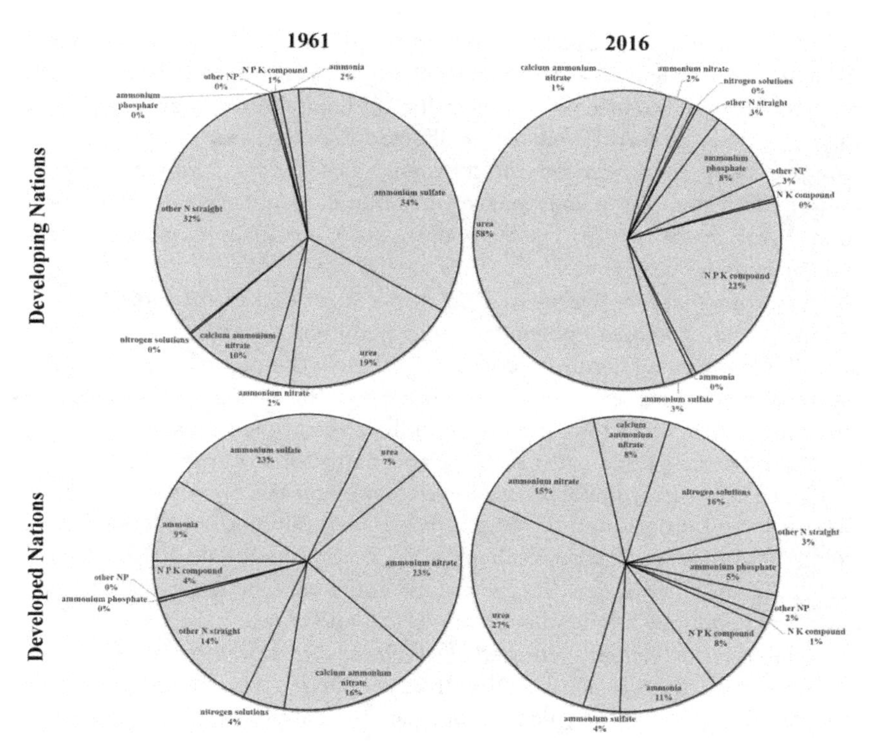

Figure 8 Global nitrogen fertilizer source consumption for developing and developed nations in 1961 and 2016 (Anon., 2019).

much research in developing more targeted in-season nitrogen applications, including slow/control-release or inhibitor-based EEFs for nitrogen.

Slow-release fertilizers involve some type of chemical or biological process. For example, urea formaldehyde, methylene urea, and triazone-based fertilizers consist of long-chain molecules containing nitrogen, which is slowly released for plant uptake as microbes decompose the molecules. These materials can be effective EEFs because they meter out the nitrogen slowly. This minimizes volatilization, denitrification, and leaching losses by avoiding a large flush of nitrogen-containing chemicals at any one time (Pierzynski et al., 2005; Bruulsema et al., 2012). These products are dependent on microbial activity and are therefore affected by factors such as very cold or hot soil temperatures. Generally, this results in a need to have an immediately available nitrogen source in the soil during cool spring conditions, especially after a fumigation. Another consideration is that these sources tend not to last the entire season, especially in warmer climates with long growing seasons. Some of these EFFs are available in liquid form and can be applied via fertigation, foliar applications, and in concentrated fluid fertilizer bands.

Sulfur coatings are used alone or in conjunction with polymer coatings. The nitrogen is released as the sulfur coating is oxidized into sulfuric acid by microbial action. Sulfur coatings have the additional advantage of releasing sulfur into the soil. The release is affected by temperature, especially in fumigated soil. Sulfur-coated products may not supply adequate nitrogen during cool parts of the season. As with other slow-release fertilizers, sulfur-coated products tend to not last the entire season, especially in warmer climates with long growing seasons.

Control-release fertilizers rely on physical processes for their mode of action. As an example, polymer-coated fertilizers (most commonly urea) absorb water through a porous coating. This swells the particle, and eventually the nutrients diffuse through the membrane as molecular diffusion speeds increase with warming temperatures and the sizes of the pores become large enough for passage due to the swelling and/or microbial degradation. The rate of release is primarily impacted by temperature and the thickness of coatings. Granules can be designed to release nutrients at differing times, ranging, for example, from 45 to 360 days (Sharma, 1979; LeMonte et al., 2016, 2018). As such, polymer-coated products can last the entire growing season if conditions are correct and they are handled carefully to avoid cracking the coatings.

Inhibitors can increase nitrogen efficiency as they slow the conversion from one form of nitrogen to another. Urease inhibitors [e.g. N-butyl-thiophosphoric triamide (NBPT)] inhibit the urease enzyme, which catalyzes the hydrolysis reaction converting urea to ammonium bicarbonate and then to ammonia gas and finally to ammonium. The ammonia gas phase renders the nitrogen very vulnerable to volatilization loss if not captured by the soil. This gas loss is nearly eliminated if the conversion from urea is slowed by use of an inhibitor, allowing the soil to capture the nitrogen more effectively. Urease inhibitors can be used with dry or fluid fertilizers. These inhibitors can be effective in all soil types, but effectiveness is especially good with soils with high pH and/or low cation-exchange capacity (CEC). They are particularly important if urea is not incorporated into the soil using tillage/injection or irrigation techniques, or in conditions which maximize losses to the atmosphere such as open crop canopies, application of liquid urea on thick crop residues or in hot, humid, and windy conditions or losses below the rooting zone due to excessive water movement through soil.

Nitrification inhibitors [e.g. Dicyandiamide (DCD), 2-chloro-6 (trichloromethyl) pyridine (nitrapyrin), 3,4-dimethylpyrazole phosphate (DMPP), and pronitridine] were developed to slow the oxidation of ammonium to nitrate by inhibiting the activity of *Nitrosomonas* spp. bacteria responsible for this conversion process. Conversion results in a molecule with a negatively charged ion that is repelled by soil and is thus subject to leaching losses, particularly with excessive precipitation/irrigation. Nitrate is also subject to gaseous loss via denitrification/nitrification. A nitrification inhibitor preserves the nitrogen in the ammonium form which minimizes the period it can be lost in its nitrate form. Their effectiveness has been evaluated by Burzaco et al. (2014). Inhibitors are especially effective in water-logged soils due to compaction and/or high clay content. They are also effective in low CEC soils prone to leaching, especially with shallow rooted crops.

It is important to understand what forms of nitrogen each inhibitor work upon. For example, urea ammonium nitrate (UAN) is the most common liquid form of nitrogen applied in many regions. Both a urease and a nitrification inhibitor can be blended with UAN. The urease inhibitor acts on only 50% of the nitrogen that is present as urea but has no impact on the ammonium and nitrate that make up the other half. The nitrification inhibitor acts on 75% of nitrogen present as ammonium (25%) and the urea once it converts to ammonium (50%). Neither inhibitor has any impact on the 25% of the nitrogen that is present as nitrate, which is at immediate risk of loss to the environment.

Normally, urea hydrolysis to ammonium is complete within 2-4 days. A urease inhibitor slows it to about 7-14 days. Conversion of ammonium to nitrate normally is complete within 7-21 days. A nitrification inhibitor slows that to about 25-55 days. Using both inhibitors extends the range to about 50-65 days. Slow-release products vary widely in their release timing, but generally are released within about 14-50 days. Because they can be more precisely engineered, polymer-coated products vary widely, depending on quality and thickness of the coating, with release timings ranging from 45 to 360 days.

One other proven EEF nitrogen product is fused ammonium nitrate with ammonium sulfate. This product was developed in response to the explosive properties of ammonium nitrate. There are many instances of fertilizer and other facilities having catastrophic explosions. This fused product greatly reduces the likelihood of this problem and, as an added benefit, provided an effective fertilizer that also contains sulfur.

The potential for improvement in nitrogen fertilization can be seen in a recent review by Omara et al. (2019) who estimated nitrogen-uptake efficiency in cereals at about 33% with some farmers achieving levels as high as 41%. These levels could be significantly improved. It has been estimated that under 10% of farmers using EEFs though studies show, for example, a near doubling in nitrogen-uptake efficiency when using products such as polymer-coated urea (LeMonte et al., 2016, 2018). Genetic developments are also needed for deeper and denser root systems that stay viable throughout the growing season to aid in taking up nitrogen.

5 Developments in phosphorus fertilizers

Phosphorus deficiency is often a 'hidden hunger' with yield losses not immediately apparent unless compared with an optimally fertilized plot (Hopkins,

2015). Figure 9 shows the major sources of phosphorus fertilizers in 1961 (near the start of the Green Revolution) and in 2016. More recent developments in phosphorus fertilizers include the development of ammoniated phosphates combining ammonium and phosphorus, including monoammonium phosphate and diammonium phosphate. A liquid form of ammoniated phosphorus has also been developed and has become the standard choice for concentrated fertilizer bands and injection into irrigation water.

These phosphorus fertilizers dissolve quickly in soils and then precipitate as iron/aluminum phosphates in acid soils and calcium/magnesium phosphates in alkaline soils. This precipitation is driven by equilibrium chemistry as the soil will only tolerate a finite concentration of phosphorus. As plants take up what phosphorus is dissolved in the soil solution, the solid-phase phosphorus precipitates re-dissolve to bring the solution back up to equilibrium. The rate of dissolution depends on the pH and the types of minerals present. In many cases, this process of dissolution is too slow to match plant requirements at differing stages of growth.

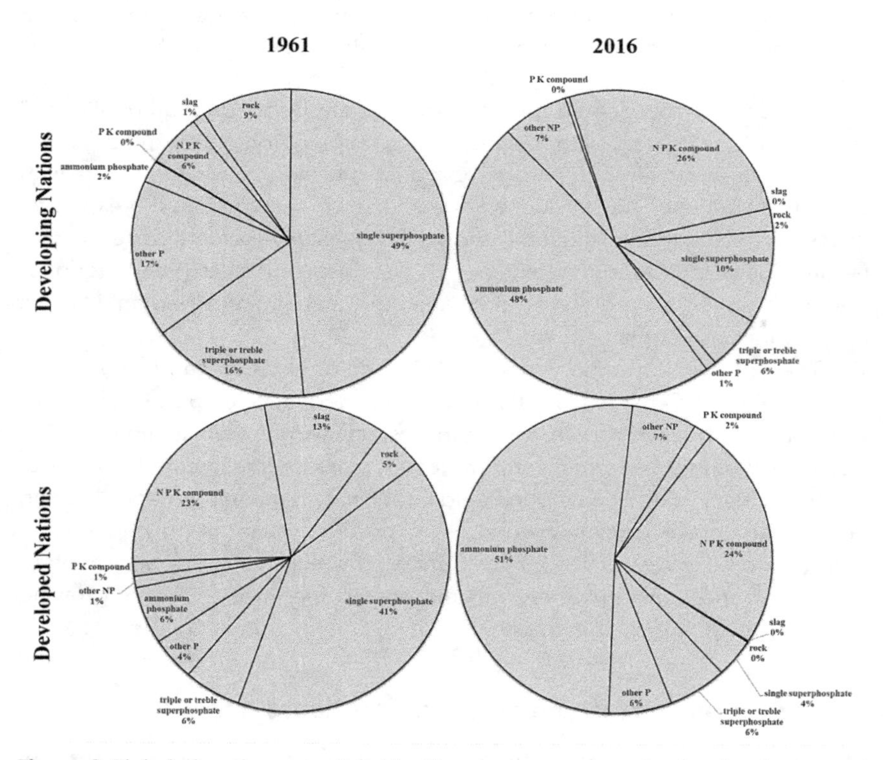

Figure 9 Global phosphorus (as P_2O_5) fertilizer source consumption for developing and developed nations in 1961 and 2016 (Anon., 2019).

Because phosphorus is less soluble than nitrogen, correct placement of fertilizer is particularly important in relation to crop root architecture and morphology (Randall and Vetsch, 2008; Hopkins and Hansen, 2019). In general, it is best to place all of the phosphorus in the root zone prior to or at planting. The deep, prolific root system of maize makes it effective in taking up phosphorus placed in the soil (Borges and Mallarino, 2001). The shallow rooting system in potato makes the uptake of phosphorus near the soil surface relatively more efficient (Hopkins and Hansen, 2019; Hopkins, 2015). The dominant taproot of sugar beet means phosphorus is best placed below the seed at planting (Anderson and Peterson, 1978; Hopkins and Hansen, 2019; Sims, 2010).

Broadcast applications of phosphorus that sit on the soil surface are relatively ineffective, especially if the soil surface is dry and/or there are few surface-feeding roots. Incorporating the fertilizer into the soil or injecting a liquid fertilizer is more effective as this will place phosphorus more directly in the path of roots. Foliar applications of phosphorus have also been shown to be effective in some cases.

As with nitrogen, there have been developments in EEF phosphorus fertilizers. Various polymer and other coatings have been developed to delay the release of phosphorus into the soil (Sharma, 1979; Nyborg et al., 1995; Yaseen et al., 2017). These can be effective as they avoid a flush of high concentration of phosphorus in soil solution followed by rapid precipitation. Instead, the phosphorus is released slowly over the season to replenish the soil solution phosphorus depleted by plant uptake.

Another development in EEFs is blending phosphorus with various organic acids (humic, fulvic, etc.; Tan, 2003; Hill et al., 2015a,b; Hopkins, 2015; Hopkins and Hansen, 2019; Hopkins et al., 2014; Olk et al., 2018; Summerhays et al., 2017). Soils in arid and semiarid regions are particularly problematic for phosphorus nutrition due to their alkaline pH and presence of high concentrations of free excess limestone (calcium carbonate, $CaCO_3$). This is particularly a problem for crops with high demand for phosphorus such as potato. Research done with these organic acids blended with ammoniated phosphorus fertilizers shows consistent increases in phosphorus uptake with associated increases in yield and crop quality in a variety of crops when grown on calcareous soils. However, research in noncalcareous soils often shows less promising results. SOM contains high concentrations of organic acids. It has been hypothesized that organic acids blended with phosphorus fertilizer are more likely to be effective when organic matter levels are low (Tan, 2003; Summerhays et al., 2015; Olk et al., 2018). Positive responses have also been reported in high organic matter soils, with researchers suggesting some type of biostimulation mechanism rather than a phosphorus response.

Another EEF phosphorus fertilizer is a maleic itaconic copolymer that is sprayed on the surface of dry phosphate fertilizers or blended with a liquid phosphate (Stark and Hopkins, 2015). The mode of action is not completely agreed upon, but a number of studies have shown it can be effective in soils with low levels of phosphorus, and with reduced rates of phosphorus fertilizer (Hopkins et al., 2018).

Another example of a phosphorus EEF is struvite (Hopkins, 2015; Hopkins and Hansen, 2019; Rech et al., 2019). Struvite is a precipitated phosphorus material that is derived from the waste streams of sewage treatment plants. Crop roots exude various organic acids, which it is thought may dissolve the struvite and result in enhanced uptake. Trials with this product have shown positive responses in various cropping systems. The use of struvite is particularly appealing because it recycles waste phosphorus and reduces the amount of mined phosphorus needed for crop production.

Another aspect of enhanced-efficiency phosphorus use is a better understanding of its interactions with other nutrients. Phosphorus and ammoniacal nitrogen applied together, especially in a concentrated liquid band, are synergistic in terms of uptake efficiency and plant growth (Hopkins, 2015). Where existing nitrogen levels are sufficient, it is important to use non-nitrogen-associated phosphorus sources. In this respect, higher-quality products with improved solubility have been developed such as potassium phosphates (Hopkins et al., 2010a). Users of phosphorus fertilizers also need to be aware that phosphorus can interact negatively with various micronutrients (especially zinc, iron, manganese, and copper; Barben et al., 2010a,b; Nichols et al., 2012).

Further work needs to be done to enhance phosphorus fertilizer efficiency. Uptake in the first year after application for a broadcast placement is only about 5–10%, although about 90% of the phosphorus is taken up after a decade (Syers et al., 2008). First-year uptake efficiency can increase to about 25–35% when placed in a concentrated band. Use of an EEF could increase efficiency by up to 50%. As with nitrogen, there is a great need to improve the genetics of crop plants in terms of improved rooting efficiency (Gill et al., 2004; Thornton et al., 2014). More roots and root hairs with greater uptake efficiency equate to better recovery of the phosphorus. This is especially true if those roots are concentrated in the surface soil where phosphorus tends to accumulate, as opposed to nitrogen that also benefits from deeper roots to recover leached nitrate.

6 Developments in potassium fertilizers

Potassium chloride (KCl; also known as muriate of potash or MOP) has been the main form of potassium fertilizer applied to crops since the advent of modern commercial fertilizer (see Fig. 10). In contrast to nitrogen and phosphorus, there has been relatively little development in potassium fertilizers. This is in part because potassium is more easily available to plants and is much less of a potential environmental risk than nitrogen or phosphorus. Unlike phosphorus, potassium is

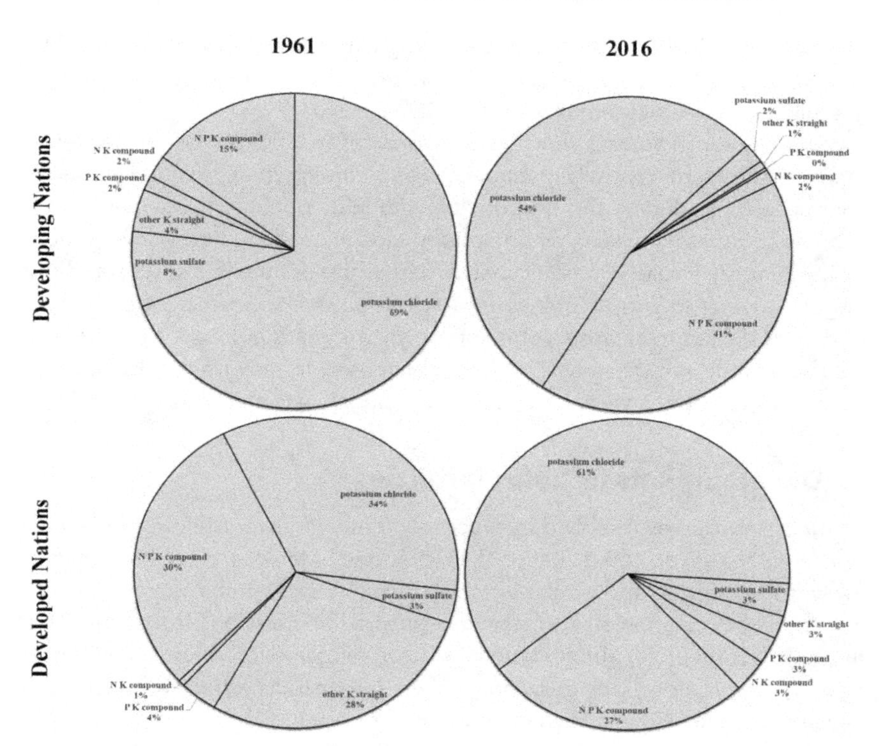

Figure 10 Global potassium (as P_2O_5) fertilizer source consumption for developing and developed nations in 1961 and 2016 (Anon., 2019).

reasonably soluble in most soils and, therefore, more readily available for uptake by plants. Unlike nitrate-N, potassium is attracted to the negatively charged soil colloids that make up a soil's CEC, which means it stays in the root zone for a reasonable amount of time after fertilization. It is susceptible to leaching losses, but much less so than nitrate, especially with relatively low CEC soils.

The second most popular form of potassium fertilizer is potassium sulfate (K_2SO_4; also known as sulfate of potash or SOP), which is made by reacting sulfuric acid with potassium chloride. This form is generally more costly but has the advantage of containing sulfur, and is preferable where both nutrients are needed and it is cost effective to do so. Other potassium products have been developed, such as potassium nitrate and potassium phosphates, which have the advantage of including other primary macronutrients.

Some soils pose problems for potassium fertilizers. Soils with few colloids (clay and organic matter), and correspondingly low CEC, do not have much ability to retain the positively charged potassium ions as efficiently. These soils, which are unusually high in sand and silt particles, also tend to have increased rate and depth of leaching. Potassium fertilizer can leach out of the root zone of low-CEC soils. Soils high in potassium-fixing clay minerals are also a problem. Some clays, such as 2:1 vermiculite clay, incorporate the potassium so tightly in the mineral that it is essentially lost for plant uptake.

In soils high in potassium-fixing clays, or with low CEC, potassium delivery to plants needs to be carefully managed through timing of application (so-called spoon feeding during the season) or with EEF potassium sources. Liquid sources of potassium, such as potassium thiosulfate, have been developed in order to apply fertilizer in liquid bands. Liquid potassium is also injected into irrigation water to supply potassium to deficient soils. Coated products have been developed to release potassium to plants gradually over the course of a season, although these are not commonly used in major agricultural crops. Foliar applications can also be effective, although costs need to be considered.

7 Developments in sulfur fertilizers

Sulfur is classified as a secondary macronutrient and historically was only rarely deficient. However, this situation has changed. Many farmers have moved away from nitrogen and potassium fertilizers that also contain sulfur. And with increasing environmental regulation to reduce sulfur emissions, less sulfur enters the soil through atmospheric deposition (acid rain) and irrigation waters. These developments, along with growing nutrient demands as yields have increased, have resulted in an increase in the frequency of sulfur deficiencies. As such, the development of sulfur fertilizers has received increasing attention. Ammonium and potassium sulfate sources are the most common forms of sulfur fertilization.

Gypsum (calcium sulfate; $CaSO_4$) is increasingly used as a form of slow-release calcium and sulfur nutrition, as well as a soil amendment. It is well proven to aid in remediating sodic soils. Elemental sulfur is also used on sodic soils, as well as with calcareous soils as a form of slow-release sulfur. This acid-forming action of sulfur can give the added benefit of solubilizing certain minerals (most notably limestone to provide calcium). Sulfur also is used with nitrogen, with elemental sulfur-coated nitrogen fertilizers developed as a slow-release form of both nitrogen and sulfur. As with nitrate-N, sulfate is an anion and repelled by soil which makes it subject to leaching. Slow-release sulfur fertilizers are used in soils particularly prone to leaching.

8 Developments in calcium, magnesium, and micronutrient fertilizers

Like sulfur, calcium and magnesium are also secondary macronutrients. However, these are generally abundant in soil and irrigation water. Like potassium, they are attracted to the negatively charged soil colloids that make up a soil's CEC and thus retained in soils. As a result of few documented deficiencies, there have been few developments in calcium and magnesium fertilizers. However, some crops, such as apple, are relatively more responsive and, thus, fertilized.

Although gypsum is increasingly used, the most common source of calcium is limestone (calcium and magnesium carbonates; $CaCO_3$ and $MgCO_3$) applied to acid soils to raise the pH–with calcium and magnesium addition as an added benefit for soils that may be deficient in these nutrients. Highly leached acid soils, especially those with low CEC, are the most likely to suffer from calcium and magnesium deficiency. In alkaline soils, limestone is not soluble and should not be used, with gypsum as a common alternative. Other sources include various calcium and magnesium nitrates, phosphates, and sulfates.

Although not as commonly deficient as the primary macronutrients, micronutrients can limit yields (Bensen et al., 2009; Hopkins et al., 1992, 2010b). This is increasingly more common for iron, zinc, manganese, copper, boron, and, to a lesser degree, chloride. This is due to increasing yields and developments in purer fertilizer products with less trace metal content. The positively charged micronutrients (e.g. zinc, manganese, and copper) are commonly available as sulfates and chlorides. These are generally effective when applied at recommended rates to soils with likely deficiencies.

Soils have about 5% total iron content. In acid soils, there is typically enough soluble iron for plant uptake. However, iron deficiencies are somewhat common in alkaline soils. The solubility of iron at high pH is near zero. Plants have evolved strategies to solubilize the iron in alkaline soils by acidifying the rhizosphere next to the root, reducing iron at the root surface, and exuding reductants or phytosiderophores (plant chelates) into the soil to make iron plant available (Hansen et al., 2006).

Scientists have synthesized chelates, such as ethylenediaminetetraacetic acid (EDTA), which help to solubilize nutrients such as calcium, zinc, and iron, although EDTA has not proved to work well in alkaline soils where the iron deficiency symptoms are most common (Hansen et al., 2006). This led to the development of ethylenediamine-N,N′-bis(2-hydroxyphenylacetic acid) (EDDHA), which can be successfully applied to soil to deliver iron into plants. This represented a significant development in combating the problem of iron-deficient crops in arid and semiarid regions. These chelates also enhance the availability of other micronutrients, such as manganese and zinc. Chelated micronutrients can be applied as soil and foliar applications. Although iron fertilizers are important, one of the greatest contributions to improved iron nutrition in plants and people was the genetic modification of rice and other plants.

Another development in EEF for micronutrients is impregnation of iron, zinc, manganese, and/or copper with elemental sulfur (Christensen et al., 2012; Hopkins et al., 2014). The elemental sulfur slowly breaks down and forms sulfuric acid, which lowers the pH of the microsite around the fertilizer prill (pellet), thus improving the solubility and uptake efficiency of the micronutrient.

9 Case study

One example of the role of fertilizers in improving crop yields is the case of Albert Huskinson, a second-generation farmer from Moody Creek, Idaho, USA. His father and uncle each received 40 acres as part of a land settlement deal and were among the first to cultivate in the Upper Snake River plain. They created a farm out of the sagebrush, raised sheep and other animals, and grew alfalfa hay, grain, and assorted garden vegetables. With the onset of flood irrigation techniques in the early twentieth century, Albert and other farmers learned that these soils and environmental conditions were very favorable to growing potatoes, particularly the Russet Burbank cultivar. Albert's yields were ~6 Mg ha^{-1}, which were similar to the global average at this time (see Fig. 3). Supplying nutrition to the potato crop was done as best as resources allowed, which involved growing alfalfa in rotation to provide nitrogen and using his livestock manure (Myers et al., 2008). His descendants, including his grandson (the author of this chapter), continue to grow this cultivar.

This scenario can be contrasted with the results of a long-term study in 2013, which found yields as high as 67 Mg ha^{-1} in well-fertilized plots (Hopkins et al., 2020). Total potato yields in the control plots, with no fertilization, were approximately one-third of the fertilized plots and, in addition, did not produce top-grade tubers which command the best prices. This can also be contrasted with average yields achieved by many smallholders in developing countries, relying essentially on sources such as legume rotation residues and animal manure for nutrition, which remain at around 7 Mg ha^{-1} in 2013 (Hopkins and Hansen, 2019).

The fertilization plan used to achieve these higher yields is shown in Table 3. This shows average rates. The growing area was divided into zones based on a range of criteria. These included physical characteristics (such as slope and aspect), soil characteristics, whether the zone had historically received applications of manure, previous yields achieved, and historical crop canopy health (based on canopy closure dates and normalized difference vegetative index or NDVI). A realistic yield goal was developed for each unique zone. Soil samples were taken in each zone. Fertilizer rates were determined for each zone and applied variably with a three-bin spreader truck with urea nitrogen and potassium chloride in separate bins, as well as a phosphorus product in a homogenous blend with sulfur, zinc, and manganese. Based on soil tests, no other nutrients were found to be deficient.

During over a century of potato farming in this area, nitrogen fertilization did have an environmental impact. The initial use of flooding irrigation leached nitrate-nitrogen into the groundwater at levels which exceeded official limits. This problem was addressed by Hopkins through the combination of EEF nitrogen sources with proper rate, timing, and placement and more efficient

Table 3 Average fertilizer application rates (kg/ha) for Russet Burbank potato trial in 2013 near Rexburg, ID

Sources	N	P_2O_5	K_2O	S	Other	Timing	Placement
Urea, polymer-coated urea (PCU), monoammonium phosphate, potassium chloride, elemental S (90%)	224 (51 urea; 173 PCU)	134	168	100	n/a	2 days prior to planting	Broadcast incorporated ~20 cm
Ammonium polyphosphate (11-37-0); chelated micronutrient blend	20	67	0	0	1 each of Zn, Mn, and Fe; 0.3 each of Cu and B; humic and fulvic acid	2 days prior to planting	7–10 cm to the side and down from seed piece
Fungicide	0	0	0	0	Trace Zn and Mn	At planting	Coated on seed pieces
Urea ammonium nitrate (32-0-0)	56	0	0	0	n/a	Split at 61 and 75 days after planting	Injected into irrigation water

Sources include percent nitrogen-phosphate-potash (N-P_2O_5-K_2O).
Rates were based on soil test and University of Idaho recommendation for Russet Burbank.

overhead sprinkler irrigation, which have virtually eliminated nitrate leaching. In addition, volatilization of ammonia and loss of gaseous nitrous oxide are nearly zero. In 2013, the cost of the fertilizer was approximately 20% of the production cost, but the cost was more than offset by increased yields and crop quality. Developments in fertilizers have enabled, along with other developments, these massive yield increases that Albert could not have achieved with the fertilizer sources and technology and knowledge he had available to him. Enabling farmers in disadvantaged areas with these is an essential part of global sustainability.

10 Conclusion and future trends

Average crop yields have more than doubled in the last six decades (Fig. 3) with nitrogen fertilizer use tripling (Fig. 11). Fertilizer use has increased dramatically in developing nations such as China, India, and Brazil. However, there are still many developing nations, largely in Africa and South America, where fertilizer availability and use is still minimal. In tracking the use of fertilizers and pesticides against increases in yields of cereals, Liu et al. (2015) identified one group of counties with both significant increases in fertilizer and pesticide consumption coinciding with increases in cereal yields, contrasting with a second group with both much lower fertilizer and pesticide use and much lower yields. They highlighted a global imbalance in food production and the (*associated*) usage of fertilizers and pesticides.

Researchers in developed nations continue to focus on increasing nutrient-uptake efficiency as they combat air- and water-quality problems.

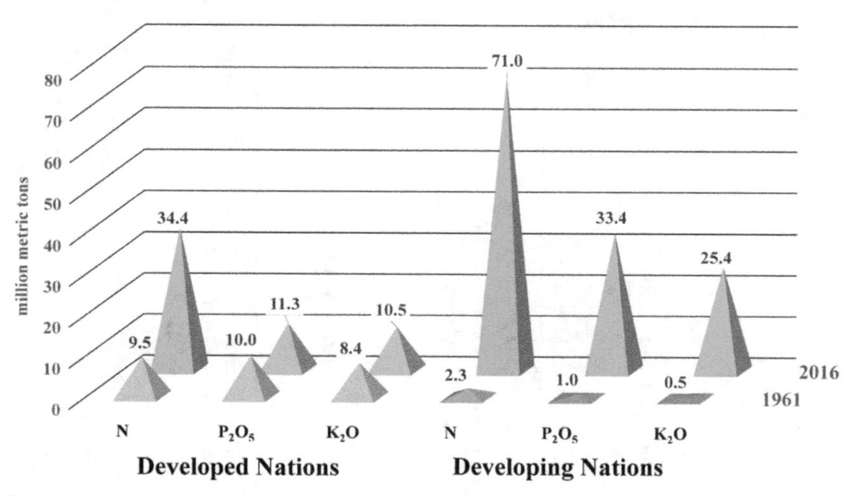

Figure 11 Primary macronutrient fertilizer (as N-P_2O_5-K_2O) consumption for 'developed' and 'developing' nations–contrasting 1961 to 2016 (Anon., 2019).

Figure 12 Enhanced efficiency fertilizers (such as the polymer-coated urea on the left) can often be applied at reduced rates than traditional fertilizers (such as the uncoated urea on the right), while simultaneously providing similar or improved yields and reduced environmental impacts.

Developments in precision agriculture, such as remote and proximal soil and crop sensors, as well as robotic delivery systems, will enable ever more targeted and precise delivery of fertilizers. Further developments in nitrogen and other inhibitors and slow/control-release products are needed (Fig. 12). Nanotechnology also provides opportunities to increase uptake efficiency. Another emerging area of research is biostimulants, with research into ways to stimulate biological nitrogen release, release of siderophores from roots and microbes, and hormonal controls to trigger plant stress responses to improve nutrient uptake. Research is also under way to transform nonlegumes, such as maize, to develop symbiotic relationships with nitrogen-fixing microbes, which will make it possible to capture nitrogen from the atmosphere in the way legumes do. Fertilizer research in agronomically disadvantaged areas will continue to focus on cultivars that can yield well with few inputs. It is vital to do more research to find methods of using inexpensive and/or locally sourced fertilizer sources (such as animal wastes and green manures).

11 Where to look for further information

Further reading

Barker, A. V. and Pillbeam, D. J. (Eds). 2015. *Handbook of Plant Nutrition*. CRC Press Taylor and Francis Group, Boca Raton, FL.

Bryson, G. M., Mills, H. A., Sasseville, D. N., Jones, J. B. and Barker, A. V. 2014. *Plant analysis Handbook IV*. MicroMacro Publishing, Athens, GA.

Engelstad, O. P. (Ed.). 1985. *Fertilizer Technology and Use* (3rd edn.). Soil Science Society of America, Madison, WI.

Emsley, J. 2000. *The 13th Element: The Sorid Tale of Murder, Fire, and Phosphorus*. John Wiley & Sons, New York, NY.

Hagar, T. 2008. *The Alchemy of Air*. Three Rivers Press, New York, NY.

Hall, A. D. 1909. *Fertilisers and Manures*. John Murray, London, UK. https://ia600308.us.archive.org/7/items/fertilisersmanur00hall/fertilisersmanur00hall.pdf (accessed on March 17, 2019).

Hignet, T. P. (Ed.). 1985. *Fertilizer Manual*. Springer Netherlands.

Lal, R. and Stewart, B. A. 2015. *Soil-specific Farming: Precision Agriculture*. CRC Press, Boca Raton, FL.

Mikkelsen, R. and Hopkins, B. G. 2009. Fertilizer BMPs – fertilizer management practices for potato production in the Pacific Northwest. *International Plant Nutrition Institute (IPNI) Special Publication* as a series of Fertilizer BMP NRCS sponsored publications. Available at: http://www.ipni.net/bmp.

Pierzynski, G. M., Vance, G. F. and Sims, J. T. 2005. *Soils and Environmental Quality*. CRC Press, Boca Raton, FL.

Rai, M. K. (Ed.). 2006. *Handbook of Microbial Biofertilizers*. Food Products Press Hawthorne Press, New York, NY.

Shannon, D. K., Clay, D. E. and Kitchen, N. R. 2018. *Precision Agriculture Basics*. American Society of Agronomy, Crop Science Society of America, and Soil Science Society of America, Madison, WI.

Key journals/conferences

- Advances in Agronomy book series
- American Society of Agronomy * Crop Science Society of America * Soil Science Society of America annual international meetings and publications
- European Confederation of Soil Science Societies annual meetings and publications
- European Journal of Soil Science
- Fertilizer Research
- Field Crops Research journal
- International Fertilizer Association meetings and publications
- International Union of Soil Sciences annual meetings and publications
- Journal of Environmental Quality
- Journal of Plant Nutrition
- Journal of Plant Nutrition and Soil Science
- Nutrient Cycling in Agroecosystems journal
- Plant and Soil journal
- Soil journal
- Soil Science Society of America Journal
- The Fertilizer Institute meetings and publications
- Western Fertilizer Handbook by California Fertilizer Association

12 References

Adisa, I. O., Pullagurala, V. L. R., Peralta-Videa, J. R., Dimkpa, C. O., Elmer, W. H., Gardea-Torresdey, J. L. and White, J. C. 2019. Recent advances in nano-enabled fertilizers and pesticides: a global review of mechanisms of action. *Environ. Sci.: Nano* 6(7), 2002–30. doi:10.1039/C9EN00265K.

Ahlgren, S., Baky, A., Bernesson, S., Nordberg, Å, Norén, O. and Hansson, P. 2012. Consequential life cycle assessment of nitrogen fertilisers based on biomass – a Swedish perspective. *Insciences J.* 2(4), 80–101. doi:10.5640/insc.020480.

Aloe, A. K., Bouraoui, F., Grizzetti, B., Bidoglio, G. and Pistocchi, A. 2014. Managing nitrogen and phosphorus loads to water bodies. JRC Technical Report 91624. Joint Research Centre of the European Commission, Luxembourg, Belgium.

Anderson, F. N. and Peterson, G. A. 1978. Optimum starter fertilizer placement for sugarbeet seedlings as determined by uptake of radioactive ^{32}P isotope. *J. Am. Soc. Sugar Beet Technol.* 20(1), 19–24. doi:10.5274/jsbr.20.1.19.

Anderson, M. S. 1960. History and development of soil testing. *J. Agric. Food Chem.* 8(2), 84–7. doi:10.1021/jf60108a001.

Andresen, M., Dresbøll, D. B., Jensen, L. S., Magid, J. and Thorup-Kristensen, K. 2016. Cultivar differences in spatial root distribution during early growth in soil, and its relation to nutrient uptake–a study of wheat, onion and lettuce. *Plant Soil* 408(1-2), 255–70. doi:10.1007/s11104-016-2932-z.

Anon. 2008. *Glossary of Soil Science Terms*. Soil Science Society of America, Madison, WI. Available at: https://www.soils.org/publications/soils-glossary (accessed 15 February 2019).

Anon. 2015. *Soil Test Levels in North America*. International Plant Nutrition Institute, Peachtree Corners, GA. Available at: http://soiltest.ipni.net/charts/change (accessed on 16 February 2019).

Anon. 2019. International fertilizer association. Available at: http://ifadata.fertilizer.org/ucResult.aspx?temp=2019081 (accessed on March 17, 2019).

Araus, J. L., Ferrio, J. P., Voltas, J., Aguilera, M. and Buxó, R. 2014. Agronomic conditions and crop evolution in ancient Near East agriculture. *Nat. Comm.* 5, 3953. doi:10.1038/ncomms4953.

Barben, S. A., Hopkins, B. G., Jolley, V. D., Webb, B. L. and Nichols, B. A. 2010a. Optimizing phosphorus and zinc concentrations in hydroponic chelator-buffered nutrient solution for Russet Burbank potato. *J. Plant Nutr.* 33(4), 557–70. doi:10.1080/01904160903506282.

Barben, S. A., Hopkins, B. G., Jolley, V. D., Webb, B. L. and Nichols, B. A. 2010b. Phosphorus and manganese interactions and their relationships with zinc in chelator-buffered solution grown Russet Burbank potato. *J. Plant Nutr.* 33(5), 752–69. doi:10.1080/01904160903575964.

Bender, R. R., Haegele, J. W., Ruffo, M. L. and Below, F. E. 2013. Nutrient uptake, partitioning, and remobilization in modern, transgenic insect-protected maize hybrids. *Agron. J.* 105(1), 161–70. doi:10.2134/agronj2012.0352.

Bensen, J. H., Geary, B. D., Miller, J. S., Jolley, V. D., Hopkins, B. G. and Stevens, M. R. 2009. *Phytophthora erythroseptica* (pink rot) development in Russet Norkotah potato grown in buffered hydroponic solutions I. Calcium nutrition effects. *Am. J. Pot. Res.* 86(6), 466–71. doi:10.1007/s12230-009-9101-3.

Berenstein, R. 2019. The use of agricultural robots in crop spraying/fertilizer applications. In: Billingsley, J. (Ed.), *Robotics and Automation for Improving Agriculture*. Burleigh Dodds Science Publishing, Cambridge, UK.

Bicer, Y., Dincer, I., Vezina, G. and Raso, F. 2017. Impact assessment and environmental evaluation of various ammonia production processes. *Environ. Manage.* 59(5), 842–55. doi:10.1007/s00267-017-0831-6.

Borges, R. and Mallarino, A. P. 2001. Deep banding phosphorus and potassium fertilizers for corn produced under ridge tillage. *Soil Sci. Soc. Am. J.* 65(2), 376–84. doi:10.2136/sssaj2001.652376x.

Bruulsema, T. W., Fixen, P. E. and Sulewski, G. D. 2012. *4R Plant Nutrition Manual: a Manual for Improving the Management of Plant Nutrition, North American Version*. International Plant Nutrition Institute, Norcross, GA.

Burzaco, J. P., Ciampitti, I. A. and Vyn, T. J. 2014. Nitrapyrin impacts on maize yield and nitrogen use efficiency with spring-applied nitrogen: field studies vs. meta-analysis comparison. *Agron. J.* 106(2), 753–60. doi:10.2134/agronj2013.0043.

Canter, L. W. 1997. *Nitrates in Groundwater*. CRC Press Taylor & Francis Group, New York, NY.

Christensen, R. C., Hopkins, B. G., Jolley, V. D., Olson, K. M., Haskell, C. M., Chariton, N. J. and Webb, B. L. 2012. Elemental sulfur impregnated with iron as a fertilizer source for Kentucky bluegrass. *J. Plant Nutr.* 35(12), 1878–95. doi:10.1080/01904167.2012.706684.

Ciampitti, I. A. and Vyn, T. J. 2014. Understanding global and historical nutrient use efficiencies for closing maize yield gaps. *Agron. J.* 106(6), 2107–17. doi:10.2134/agronj14.0025.

Cocker, D., Orris, G. and Wynee, J. 2016. US geological survey assessment of global potash production and resources. Special Paper 520. The Geological Society of America, Washington DC.

Cole, D. L., Woolley, R. K., Tyler, A., Buck, R. L. and Hopkins, B. G. 2020. Mineral nutrient deficiencies in quinoa grown in hydroponics with single nutrient salt/acid/chelate sources. *J. Plant Nutr.* 43: (accepted for publication).

Dordas, C. 2008. Role of in controlling plant diseases in sustainable agriculture. A review. *Agron. Sustain. Dev.* 28(1), 33–46. doi:10.1051/agro:2007051.

Fageria, N. K., Barbosa Filho, M. P. B., Moreira, A. and Guimarães, C. M. 2009. Foliar fertilization of crop plants. *J. Plant Nutr.* 32(6), 1044–64. doi:10.1080/01904160902872826.

Fernelius, K. J., Madsen, M. D., Hopkins, B. G., Bansal, S., Anderson, V. J., Eggett, D. L. and Roundy, B. A. 2017. Post-fire interactions between soil water repellency, soil fertility and plant growth in soil collected from a burned piñon-juniper woodland. *J. Arid Env.* 144, 98–109. doi:10.1016/j.jaridenv.2017.04.005.

Geary, B. D., Clark, J., Hopkins, B. G. and Jolley, V. D. 2015. Deficient, adequate and excess nitrogen levels established in hydroponics for biotic and abiotic stress-interaction studies in potato. *J. Plant Nutr.* 38(1), 41–50. doi:10.1080/01904167.2014.912323.

Gill, H. S., Singh, A., Sethi, S. K. and Behl, R. K. 2004. Phosphorus uptake and use efficiency in different varieties of bread wheat (*Triticum aestivum* L). *Arch. Agron. Soil Sci.* 50(6), 563–72. doi:10.1080/03650340410001729708.

Guthrie, S., Giles, S., Dunkerley, F., Tabaqchali, H., Harshfield, A., Ioppolo, B. and Manville, C. 2018. *The Impact of Ammonia Emissions from Agriculture on Biodiversity: an Evidence Synthesis*. The RAND Corporation/The Royal Society, London, UK.

Hansen, N. C., Hopkins, B. G., Ellsworth, J. W. and Jolley, V. D. 2006. Iron nutrition in field crops. In: Barton, L. L. and Abadia, J. (Eds), *Iron Nutrition in Plants and Rhizospheric Microorganisms*. Springer Publishing, New York, NY, pp. 21–53.

Hasler, K., Broring, S., Omta, S. W. F. and Olfs, H.-W. 2015. Life cycle assessment (LCA) of different fertilizer product types. *Eur. J. Agron.* 69, 41–51. doi:10.1016/j.eja.2015.06.001.

Hill, M. W., Hopkins, B. G. and Jolley, V. D. 2015a. Maize in-season growth response to organic acid-bonded phosphorus fertilizer (Carbond P®). *J. Plant Nutr.* 38(9), 1398–415. doi:10.1080/01904167.2014.973040.

Hill, M. W., Hopkins, B. G., Jolley, V. D. and Webb, B. L. 2015b. Phosphorus mobility through soil increased with organic acid-bonded phosphorus fertilizer (Carbond® P). *J. Plant Nutr.* 38(9), 1416–26. doi:10.1080/01904167.2014.973041.

Hopkins, B. G. 2015. Phosphorus in plant nutrition. In: Pilbeam, D. J. and Barker, A. V. (Eds), *Plant Nutrition Handbook* (2nd edn.). CRC Press, Taylor & Francis Group, Boca Raton, FL, pp. 65–126. Chapter 3.

Hopkins, B. G. and Hansen, N. C. 2019. Phosphorus management in high yield systems. *J. Environ. Qual.* 48(5), 1265–80. doi:10.2134/jeq2019.03.0130.

Hopkins, B. G. and Hirnyck, R. E. 2007. Organic potato production. In: Johnson, D. A. (Ed.), *Potato Health Management*. American Phytopathological Society, Minneapolis, MN, pp. 101–8. Chapter 11.

Hopkins, B. G., Jolley, V. D. and Brown, J. C. 1992. Differential response of Fe-inefficient muskmelon, tomato, and soybean to phytosiderophore released by Coker 227 oat. *J. Plant Nutr.* 15(1), 35–48. doi:10.1080/01904169209364300.

Hopkins, B. G., Rosen, C. J., Shiffler, A. K. and Taysom, T. W. 2008. Enhanced efficiency fertilizers for improved nutrient management: potato (*Solanum tuberosum*). *Crop Manag.* (online). doi:10.1094/CM-2008-0317-01-RV.

Hopkins, B. G., Ellsworth, A. G., Shiffler, A. K., Cook, A. G. and Bowen, T. R. 2010a. Monopotassium phosphate as an in-season fertigation option for potato. *J. Plant Nutr.* 33(10), 1422–34. doi:10.1080/01904167.2010.489981.

Hopkins, B. G., Jolley, V. D., Webb, B. L. and Callahan, R. K. 2010b. Boron fertilization and evaluation of four soil extractants: russet Burbank potato. *Commun. Soil Sci. Plant Anal.* 41(5), 527–39. doi:10.1080/00103620903527928.

Hopkins, B. G., Horneck, D. A. and MacGuidwin, A. E. 2014. Improving phosphorus use efficiency through potato rhizosphere modification and extension. *Am. J. Potato Res.* 91(2), 161–74. doi:10.1007/s12230-014-9370-3.

Hopkins, B. G., Fernelius, K. J., Hansen, N. C. and Eggett, D. L. 2018. AVAIL phosphorus fertilizer enhancer: meta-analysis of 503 field evaluations. *Agron. J.* 110(1), 389–98. doi:10.2134/agronj2017.07.0385.

Hopkins, B. G., Stark, J. C. and Kelling, K. A. 2020. Nutrient management. In: Stark, J. C., Thornton, M. K. and Nolte, P. (Eds), *Potato Production Systems*. Springer Publishing, New York (in press).

Kibblewhite, M. 2007. Implications of farm management on the nutrient cycle. In: Hislop, H. (Ed.), *The Nutrient Cycle: Closing the Loop*. Green Alliance, London, UK.

Lal, R. and Stewart, B. A. 2015. *Soil-specific Farming: Precision Agriculture*. CRC Press, Boca Raton, FL.

LeMonte, J. J., Jolley, V. D., Summerhays, J. S. C., Terry, R. E. and Hopkins, B. G. 2016. Polymer coated urea in turfgrass maintains vigor and mitigates nitrogen's environmental impacts. *PLoS ONE* 11(1), e0146761. doi:10.1371/journal.pone.0146761.

LeMonte, J. J., Jolley, V. D., Story, T. M. and Hopkins, B. G. 2018. Assessing atmospheric nitrogen losses with photoacoustic infrared spectroscopy: polymer coated urea. *PLoS ONE* 13(9), e0204090. doi:10.1371/journal.pone.0204090.

Li, T., Zheng, W., Yin, J., Chadwick, D., Norse, D., Lu, Y., Liu, X., Chen, X., Zhang, F., Powlson, D. and Dou, Z. 2018. Enhanced-efficiency fertilizers are not a panacea for resolving the nitrogen problem. *Glob. Change Biol.* 24(2), e511–21. doi:10.1111/gcb.13918.

Liu, Y., Pan, X. and Li, J. 2015. A 1961–2010 record of fertilizer use, pesticide application and cereal yields: a review. *Agron. Sustain. Dev.* 35(1), 83–93. doi:10.1007/s13593-014-0259-9.

Long, D. 2018. Site-specific nutrient-management systems. In: Stafford, J. (Ed.), *Precision Agriculture for Sustainability*. Burleigh Dodds Science Publishing, Cambridge, UK.

Mallarino, A. P. 2003. Field calibration for corn of the Mehlich-3 soil phosphorus test with colorimetric and inductively coupled plasma emission spectroscopy determination methods. *Soil Sci. Soc. Am. J.* 67(6), 1928–34. doi:10.2136/sssaj2003.1928.

Marschner, P. and Rengel, Z. 2007. *Nutrient Cycling in Terrestrial Ecosystems*. Springer Publishers, Heidelberg, Germany.

Motavalli, P. P., Goyne, K. W. and Udavatta, R. P. 2008. Environmental impact of enhanced-efficiency nitrogen fertilizers. *Crop Man.* 7(1), 1–10. doi:10.1094/CM-2008-0730-02-RV.

Myers, P., McIntosh, C. S., Patterson, P. E., Taylor, R. G. and Hopkins, B. G. 2008. Optimal crop rotation of Idaho potatoes. *Am. J. Pot Res* 85(3), 183–97. doi:10.1007/s12230-008-9026-2.

Nichols, B. A., Hopkins, B. G., Jolley, V. D., Webb, B. L., Greenwood, B. G. and Buck, J. R. 2012. Phosphorus and zinc interactions and their relationships with other nutrients in maize grown in chelator-buffered nutrient solution. *J. Plant Nutr.* 35(1), 123–41. doi:10.1080/01904167.2012.631672.

Nyborg, M., Solberg, E. D. and Pauly, D. G. 1995. Coating of phosphorus fertilizers with polymers increases crop yield and fertilizer efficiency. *Better Crops Plant Food* 79(3), 8–9. Available at: http://www.ipni.net/publication/bettercrops.nsf/0/FF77F99FAEDBEBC385257D2E006EDD60/$FILE/BC-1995-3%20p8.pdf (accessed on 17 March 2019).

Olk, D. C., Dinnes, D. L., Scoresby, J. R., Callaway, C. R. and Darlington, J. W. 2018. Humic products in agriculture: potential benefits and research challenges–a review. *J. Soils Sediments* 18(8), 2881–91. doi:10.1007/s11368-018-1916-4.

Omara, P., Aula, L., Oyebiyi, F. and Raun, W. R. 2019. World cereal nitrogen use efficiency trends: review and current knowledge. *Agrosyst. Geosci. Environ.* 2(1). doi:10.2134/age2018.10.0045.

Parkin, T. B. and Hatfield, J. L. 2014. Enhanced efficiency fertilizers: effect on nitrous oxide emissions in Iowa. *Agron. J.* 106(2), 694–702. doi:10.2134/agronj2013.0219.

Peck, T. R. 1990. Soil testing past, present, and future. *Commun. Soil Sci. Plant Anal.* 21(13-16), 1165–86. doi:10.1080/00103629009368297.

Pierzynski, G. M., Vance, G. F. and Sims, T. J. 2005. *Soils and Environmental Quality* (3rd edn.). CRC Press Taylor & Francis Group, Boca Raton, FL.

Randall, G. and Vetsch, J. 2008. Optimum placement of phosphorus for corn/soybean rotations in a strip-tillage system. *J. Soil Water Cons.* 63(5), 152A–3A. doi:10.2489/jswc.63.5.152A.

Rech, I., Withers, P., Jones, D. and Pavinato, P. 2019. Solubility, diffusion and crop uptake of phosphorus in three different struvites. *Sustainability* 11(1), 134. doi:10.3390/su11010134.

Ritchie, S. W., Hanway, J. J. and Benson, G. O. 1997. How a corn plant develops. Spec. Publ. 48. Iowa State Univ. Coop. Ext. Serv., Ames, IA.

Schlegel, A. J. and Havlin, J. L. 2017. Corn yield and grain nutrient uptake from 50 years of nitrogen and phosphorus fertilization. *Agron. J.* 109(1), 335–42. doi:10.2134/agronj2016.05.0294.

Sharma, G. C. 1979. Controlled-release fertilizers and horticultural applications. *Sci. Hortic.* 11(2), 107–29. doi:10.1016/0304-4238(79)90037-2.

Sharpley, A., Jarvie, H., Flaten, D. and Kleinman, P. 2018. Celebrating the 350th anniversary of phosphorus discovery: a conundrum of deficiency and excess. *J. Environ. Qual.* 47(4), 774–7. doi:10.2134/jeq2018.05.0170.

Sims, A. L. 2010. Sugarbeet response to broadcast and starter phosphorus applications in the Red River Valley of Minnesota. *Agron. J.* 102(5), 1369–78. doi:10.2134/agronj2010.0099.

Skowronska, M. and Filipek, T. 2014. Life cycle assessment of fertilizers: a review. *Int. Agrophys.* 28(1), 101–10. doi:10.2478/intag-2013-0032.

Snyder, C. S., Bruulsema, T. W., Jensen, T. L. and Fixen, P. E. 2009. Review of greenhouse gas emissions from crop production systems and fertilizer management effects. *Agric. Ecosyst. Environ.* 133(3-4), 247–66. doi:10.1016/j.agee.2009.04.021.

Stafford, J. (Ed.) 2018. *Precision Agriculture for Sustainability*. Burleigh Dodds Science Publishing, Cambridge, UK.

Stark, J. C. and Hopkins, B. G. 2015. Fall and spring phosphorus fertilization of potato using a dicarboxylic acid polymer (AVAIL®). *J. Plant Nutr.* 38(10), 1595–610 (online first). doi:10.1080/01904167.2014.983124.

Summerhays, J. S. C., Hopkins, B. G., Jolley, V. D., Hill, M. W., Ransom, C. J. and Brown, T. R. 2015. Enhanced phosphorus fertilizer (Carbond P®) supplied to maize in moderate and high organic matter soils. *J. Plant Nutr.* 38(9), 1359–71. doi:10.1080/01904167.2014.973039.

Summerhays, J. S. C., Jolley, V. D., Hill, M. W. and Hopkins, B. G. 2017. Enhanced phosphorus fertilizers (Carbond P® and AVAIL®) supplied to maize in hydroponics. *J. Plant Nutr.* 40(20), 2889–97. doi:10.1080/01904167.2017.1384007.

Syers, J. K., Johnston, A. E. and Curtin, D. 2008. FAO fertilizer and plant nutrition bulletin 18. Efficiency of soil and fertilizer phosphorus use. Food and Agriculture Organization of the United Nations, Rome. Available at: http://www.fao.org/3/a1595e/a1595e00.htm (accessed on 16 February 2019).

Tan, K. H. 2003. *Humic Matter in Soil and the Environment: Principles and Controversies*. Marcel Dekker Inc, New York, NY.

Thornton, M. K., Novy, R. G. and Stark, J. C. 2014. Improving phosphorus use efficiency in the future. *Am. J. Potato Res.* 91(2), 175–9. doi:10.1007/s12230-014-9369-9.

Van Groenigen, J. W., Velthof, G. L., Oenema, O., Van Groenigen, K. J. and Van Kessel, C. 2010. Towards an agronomic assessment of N_2O emissions: a case study for arable crops. *Eur. J. Soil Sci.* 61(6), 903–13. doi:10.1111/j.1365-2389.2009.01217.x.

Venterea, R. T., Hyatt, C. R. and Rosen, C. J. 2011. Fertilizer management effects on nitrate leaching and indirect nitrous oxide emissions in irrigated potato production. *J. Environ. Qual.* 40(4), 1103–12. doi:10.2134/jeq2010.0540.

Venterea, R. T., Coulter, J. A. and Dolan, M. S. 2016. Evaluation of intensive "4R" strategies for decreasing nitrous oxide emissions and nitrogen surplus in rainfed corn. *J. Environ. Qual.* 45(4), 1186–95. doi:10.2134/jeq2016.01.0024.

Yaseen, M., Aziz, M. Z., Manzoor, A., Naveed, M., Hamid, Y., Noor, S. and Khalid, M. A. 2017. Promoting growth, yield, and phosphorus-use efficiency of crops in maize–wheat cropping system by using polymer-coated diammonium phosphate. *Comm. Soil Sci. Plant Anal.* 48(6), 646–55. doi:10.1080/00103624.2017.1282510.

Zhang, X., Davidson, E. A., Mauzerall, D. L., Searchinger, T. D., Dumas, P. and Shen, Y. 2015. Managing nitrogen for sustainable development. *Nature* 528(7580), 51-9. doi:10.1038/nature15743.

Chapter 4

Advances in fertigation techniques to optimize crop nutrition

Asher Bar-Tal, Uri Yermiyahu and Alon Ben-Gal, Agricultural Research Organization (ARO), Israel

1 Introduction

Sustainable agriculture requires integrated systems of plant and animal production to allow the supply of current food and fiber needs without compromising the future ability to continue such a supply. From the point of view of the soil, the goal is to produce food while maintaining fertility for future generations and avoiding environmental pollution. Common approaches to achieve these goals are through low-input agriculture, minimizing the use of synthetic fertilizers and recycling of organic wastes and plant residues as nutrient sources. However, it is important to keep in mind that 'sustainable' is not synonymous with 'organic' or 'low-input agriculture'. Actually, intensive organic agriculture relying on organic matter, such as composted manure added to the soil prior to planting, has been shown to result in significant leaching of nitrates through the vadose zone to groundwater, whereas similar intensive conventional agriculture implementing liquid fertilizer through drip irrigation resulted in lower rates of pollution of the vadose zone and groundwater (Dahan et al., 2014). Accurate fertilization methods that apply

http://dx.doi.org/10.19103/AS.2019.0062.28

and distribute fertilizers through irrigation systems (according to plant demand) during the growing season dramatically reduce the potential for groundwater contamination from both organic and conventional greenhouses (Dahan et al., 2014; Yasuor et al., 2013).

Within the realm of modern agriculture, sustainability refers to how well inputs and processes maintain an adequate level of production to meet demand while preserving the integrity of the environment and protecting consumers. On the one hand, this is commonly driven by a minimalist approach of using only what is needed to reduce waste. On the other hand, to meet increased demand for food spurred by a larger and richer population, the Food and Agriculture Organization (FAO) has projected that the global agricultural production in 2050 will be 60% higher than that in 2005–2007. Most of this necessary increase in production over the next 40 years is expected to be derived from improved yields (Alexandratos and Bruinsma, 2012; FAO, 2017).

The United States Department of Agriculture (USDA) and the International Plant Nutrition Institute (IPNI) promoted a philosophy of farmer nutrient management based on the following four pillars: (1) the right rate (match the amount of fertilizer to crop needs); (2) the right time (make nutrients available when the crop needs them); (3) the right place (keep nutrients where crops can use them) and (4) the right source/kind of fertilizer (matched to crop needs). The National Institute of Food and Agriculture (NIFA) of the USDA declared that existing and future NIFA-supported research, coupled with the application of this management philosophy, will promote sustainable use of nutrients (Elliott, 2016).

In this chapter, we describe fertigation as a method that meets the philosophy described earlier and contributes to sustainable management of plant nutrition. We continue with a number of case studies and examples and finish with thoughts concerning the future of fertigation, including research needs and trends.

2 The right rate at the right time: nutrient consumption curves and supply

The right rate and time of fertilization of any crop is determined by the amounts of nutrients required at specific physiological stages for optimal yield. The optimal uptake of nutrients varies widely during crop growth. Specific environmental conditions, climate, CO_2 concentration, water supply, soil properties and nutrient availability all can strongly impact nutrient requirements. The consumption curves under different environmental conditions are controlled by these factors and their interactions, and therefore optimal fertigation management is a challenge. The empirical consumption curves, which are the quantitative description of the optimal uptake of nutrients as a function of time, are useful for understanding and managing fertilizer requirements. The nutrient consumption curves and the dry matter production curves are correlated to each other, but there are differences between the two that vary with the developmental stage and between specific nutrients. There are considerable differences among crops and among varieties of the same species in the uptake rate and the time at which the maximum consumption rate occurs. The consumption of nutrients by many crops can change sharply as a function of physiological stages of plant development. At each growth phase the rate of nutrient requirement is determined by two main processes: (i) formation of new vegetative plant tissues and (ii) formation of reproductive organs (flowers, fruits, seeds, etc.) (Bar-Yosef, 1999).

Daily nutrient uptake rates are derived from consumption curves, and those that result in optimum yield and product quality are crop specific and depend on weather conditions. The lack of attention to changes in the uptake rate over time may lead to periods of over- or under-fertilization. Over-fertilization may exacerbate soil salinity and environmental contamination, whereas under-fertilization may result in nutrient deficiency and yield reduction (Bar-Yosef, 1999). The rate of nutrient uptake by a leafy vegetable (lettuce, *Lactuca sativa*) was found to be characterized by an exponential curve, increasing sharply over time (Silber et al., 2003). The general nutrient consumption curve of fruit-bearing crops has been characterized by three periods: an exponential rate during the initial vegetative growth, followed by a linear growth rate, and finally a declining rate as reproductive organs develop (Hochmuth, 1992). This consumption curve closely fits published measurements of nutrient uptake by determinant crops such as corn (*Zea mays* L.) and wheat (*Triticum aestivum*) (Bar-Tal et al., 2004; Beraud et al., 2005) and fruiting vegetables such as topped tomatoes (*Solanum lycopersicum*) (Tanaka et al., 1974). When non-determinant fruits such as tomatoes and peppers (*Capsicum annuum*) were grown continuously under well-controlled climatic conditions, their nutrient uptake rate grew steadily until the production of the first fruit truss and then became monotonic (Bar-Yosef, 1999; Bar-Tal et al., 2001; Gallardo et al., 2009, 2014).

The consumption curves for N, P and K should be used cautiously and treated only as a first approximation when applied to different environmental conditions (Bar-Yosef, 1999). For example, pepper (*C. annuum*) plants irrigated with the same N concentration took up 2.2–2.8 times higher N in the summer compared to winter (Xu et al., 2001). Not only does the total demand of nutrients fluctuate with time, but also the specific demand for individual nutrients can vary. For example, the accumulated transpiration and consumption curves of N, K and Ca by pepper plants were found to differ as a function of N level (Fig. 1). One of the major advantages of fertigation is that the amounts and concentrations of specific nutrients can be adjusted to crop requirements according to the stage of development and environmental conditions. Although the growth and

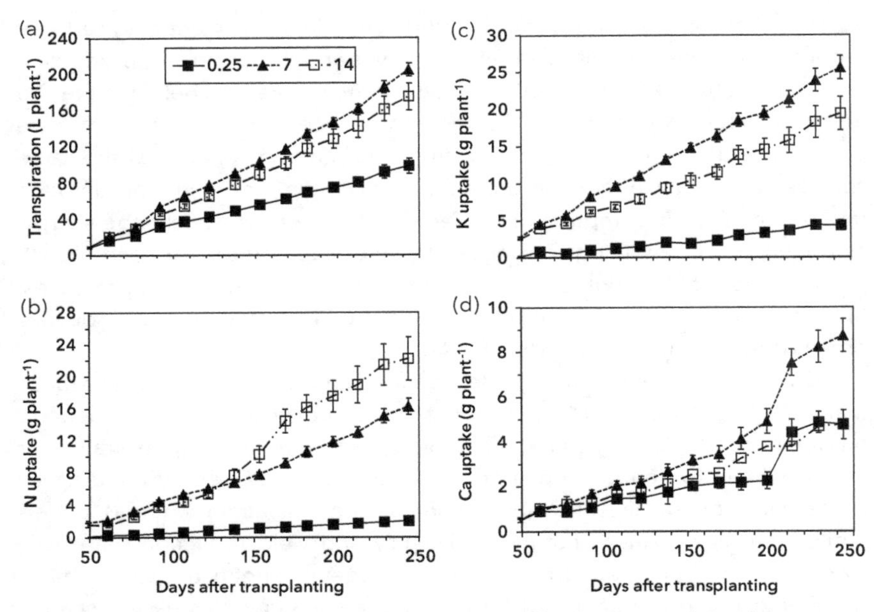

Figure 1 Transpiration (a), and consumption curves of N (b), K (c) and Ca (d) by pepper plant with time at different N dose treatments (0.25, 7.0, 14.0 mmol N/L in solution). Means ± SE. Source: adapted from Bar-Tal et al. (2001).

transpiration rates have a strong impact on the global uptake of nutrients, the uptake of individual nutrients is influenced by the physiological stage of growth (Bar-Tal et al., 2001; Voogt and Bar-Yosef, 2019).

3 Irrigation frequency

Irrigation frequency is an important management factor to consider in supplying water and nutrients. A major advantage of micro-irrigation over other methods lies in the potential for highly frequent applications. The beneficial effects of high-frequency irrigation were recognized some decades ago, and it is considered a useful tool for optimizing the root environment (Rawlins and Raats, 1975; Bar-Yosef, 1999; Segal et al., 2006). One of the problems in reviewing this issue is in the definition of 'high-frequency' irrigation. Even with drip systems, daily irrigation can be found at either end of the frequency scale, depending on the specific crop, soil and climate. For cool-climate crops in heavy soils, irrigation once every several weeks may be the norm and daily irrigation may be considered extremely frequent (Lamm and Trooien, 2003; Kang et al., 2004; El-Hawary, 2005). In contrast, irrigation of fruit or vegetable crops in light soils or soilless media often can include multiple short pulses per day or continuous, low-flow water application throughout the hours of evapotranspiration (Assouline, 2002; Ben-Gal and Dudley, 2003; Assouline et al., 2006; Segal et al., 2006). In dry, hot climates such pulsed or continuous irrigation regimes are becoming increasingly popular.

Positive effects on plant growth under frequent drip irrigation have been attributed to near-constant conditions in the root zone, allowing plants to grow roots in areas with favorable water, oxygen, nutrient and salt concentrations (Rawlins and Raats, 1975; Bravdo et al., 1992; Clothier and Green, 1994). Thus, the uptake of water and nutrients is optimized, as the replenishment of soil water adjacent to active roots is maximal and constant. As a result of increasing irrigation frequency, advantageous conditions for plant growth are exploited, and vegetative and reproductive growth benefits alternatively from the conditions or the lower energy allocated to growing new roots to find water and nutrients.

When ions are supplied to the soil with water, their concentrations in the soil solution decrease with time due to their adsorption onto solid phases and precipitation of insoluble compounds. Therefore, high concentrations of nutrients used in fertigation with low-frequency irrigation lead to fluctuations in the rhizosphere from high, or even excessive, concentrations, immediately after irrigation, to potentially deficit levels as time proceeds (Fig. 2a).

Reducing the time interval between successive irrigations may reduce variations in nutrient concentration and enable fertigation at concentrations required by the roots (Fig. 2b). Frequent replenishment of nutrients minimizes the depletion zone formed at the root surface caused by the uptake of nutrients during the period between successive irrigation events, decreases the concentration gradient between the medium solution and the root interface and diminishes the role of diffusion in nutrient transport toward the roots (Bar-Tal et al., 1994; Ben-Gal and Dudley, 2003). The impact of fertigation frequency on the uptake of macronutrient elements by plants follows the expected order of P > K > N (Kargbo et al., 1991; Silber et al., 2003; Xu et al., 2004). However, some studies showed that the frequency of injection of liquid N fertilizer into the irrigating water had no effect on crop yield and nitrogen uptake (Locascio et al., 1997; Neary et al., 1995; Thompson et al., 2003).

Irrigation frequency influences root system architecture and root length through two main mechanisms: (i) a direct effect on the wetting patterns and water distribution in the soil volume, both of which modulate root distribution and growth (Phene et al., 1991; Coelho and Or, 1997); and (ii) an indirect effect on nutrient availability (Lorenzo and Forde, 2001), especially that of P (Ben-Gal and Dudley, 2003), which significantly modifies root system efficiency (Lynch et al., 2005), including root hair density (Ma et al., 2001) and root system architecture (Williamson et al., 2001).

Yield gains under high irrigation frequency have been credited alternatively to the availability and uptake efficiency of water (Segal et al., 2006) or increased

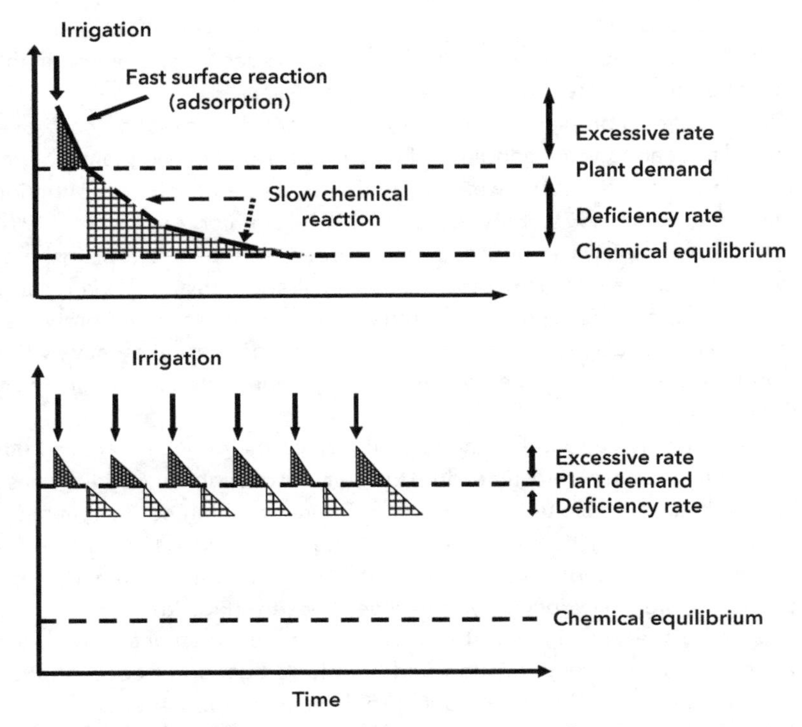

Figure 2 Schematic representation of the time variation of nutrient concentrations in the rhizosphere under conventional (a) and frequent irrigation (b). Arrows indicate one irrigation event in (a) and five irrigation events in (b) at the same time range. Excess and deficiency rates correspond to nutrient concentrations above or below plant demand, respectively; chemical equilibrium corresponds to nutrient concentrations governed by equilibrium processes in the soil.

availability of nutrients, especially P (Silber et al., 2003; Ben-Gal and Dudley, 2003). We suggest that, even though P mobilization and uptake are certainly major contributors to the benefits of high-frequency drip irrigation, this cannot and should not be un-coupled from water and increases in the efficiency of its uptake.

4 The right place: water, salt, nutrient and root development under irrigation

The aim of fertigation is to supply nutrients into the wetted soil where crop root systems are concentrated and active. As described in the following text, the transport and distribution of some nutrients (such as nitrogen as nitrate) under fertigation may follow water distribution. In such a case, fertigation optimally localizes nitrate where roots are most active and the uptake is most efficient. Recently, it was shown that fertigation with drip and micro-sprinklers increased grain yield and nitrogen uptake of winter wheat (*Triticum aestivum*) and reduced nitrate leaching below the root zone when compared with traditional fertilization and flood irrigation (Li et al., 2018, 2019). The higher water-use efficiency (WUE) and nitrogen-use efficiency (NUE) in the micro-irrigation treatments were related to the co-location of roots, water and N-fertilizer distribution. Yasuor et al. (2013) demonstrated that using fertigation in soil-grown pepper (*C. annuum*) under commercial conditions enabled high NUE (90%) without negatively affecting fruit yield or quality and minimized environmental pollution by nitrogen.

The distribution of water in the soil irrigated by a point source is governed by soil properties and the discharge rate of drippers. The movement of water in the soil from a dripper is driven by capillary and gravity forces. This creates a wetted soil volume with the soil varying in moisture content over depth (Bresler, 1978). The general geometry of the wetted soil volume below and around an emitter of a surface drip system is a symmetrical onion shape with the highest moisture close to the emitter and soil surface and a gradual decrease with the vertical and horizontal distance from the emitter down to a sharp wetting front. In light-textured soil with high hydraulic conductivity, the depth is greater and the horizontal distance is shorter than in heavier-textured soil with lower hydraulic conductivity. Most often, water movement from a point source is described as three-dimensional infiltration followed by redistribution in the soil using Richards' equation (Molz, 1981). Solute transport in the soil is commonly described by the convention-dispersion equation, taking into account convection, diffusion and dispersion. In convective transport, solutes are carried by the mass flow of water. Diffusive transport describes the diffusion of solutes from an area of high concentration to that of low concentration. Dispersion is due to the mixing of solutes as a function of variable pore velocities and differential transport between pores of varied size and shape in the soil. Consequently, soluble ions and molecules with neutral charge move from the dripper toward the borders of the wetting front as shown in Fig. 3. However, differences in the transport and distribution of nutrients and water in the soil under fertigation occur due to the injection of fertilizers during some of the irrigation events (Ben-Gal and Dudley, 2003). Soil type and discharge rate of emitters influenced the distribution of volumetric salt content (Bresler, 1978; Bar-Yosef and Sheikholslami, 1976). Under identical discharge rates, the lateral movement of salts in a sandy soil was about half that of a loam soil, and the downward movement was three times greater. The transport of ions is also affected by soil properties and the affinity of ions to soil particle surfaces, leading to differences in the distribution of various ions and water under drip irrigation (Ben-Gal and Dudley, 2003; Palacios-Díaz et al., 2009; Hassan et al., 2010).

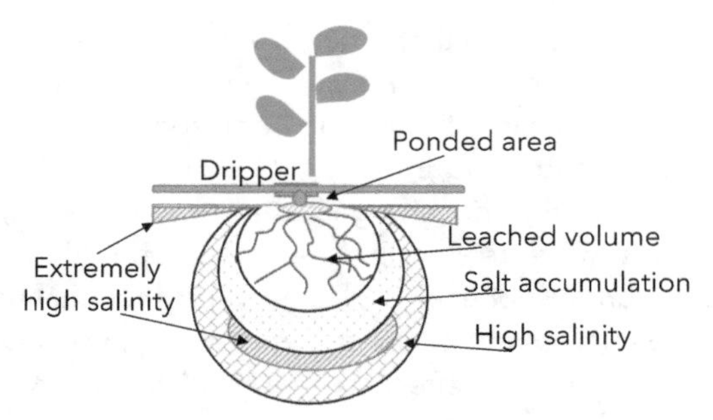

Figure 3 Schematic representation of salt distribution in the wetted soil volume below an emitter. Source: based on Kremmer and Kenig (1996).

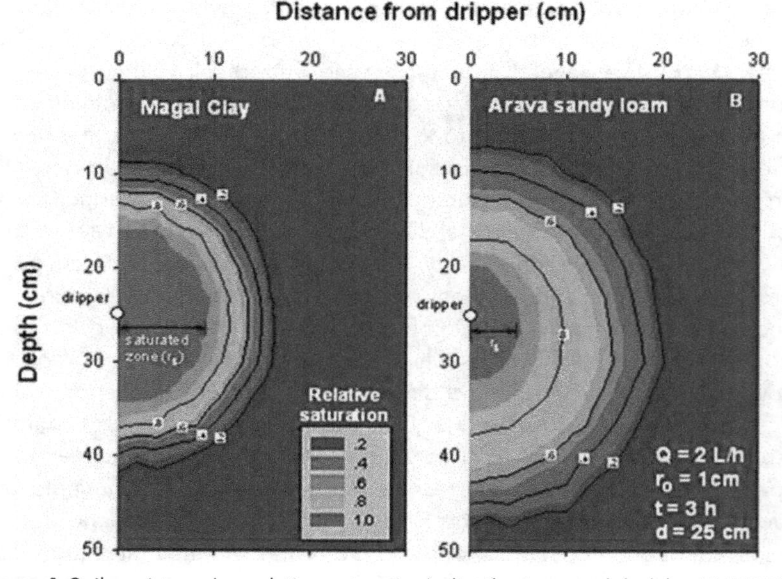

Figure 4 Soil moisture (as relative saturation) distribution modeled by HYDRUS-2D for two soils under subsurface drip irrigation. Q = dripper flow rate, r_0 = cavity radius, t = time, d = depth, r_s = saturated radius. Kindly provided by Naftali Lazarovitch; for the references see Ben-Gal and Lazarovitch (2003) and Ben-Gal et al. (2004).

When subsurface drip irrigation is practiced, the wetted volume is centered on an emitter; the water above and below the emitter is unevenly distributed, with longer vertical distance below the emitter than above it (Fig. 4). Root water uptake is related to root density and reportedly varies nonlinearly with depth down the soil profile (Chandra and Rai, 1996; Hayhoe, 1981). If soils are irrigated frequently, especially from the surface, they remain relatively wet there and most of the root water uptake takes place in the upper soil layers (Klepper,

1991). Coelho and Or (1997) characterized two-dimensional root distribution for drip-irrigated corn plants. They fitted Gaussian distribution parametric models to the corn root length density (RLD) to produce two-dimensional root distributions that they compared to root water uptake (RWU) patterns as shown in Fig. 5. Although actual water uptake patterns are a result of complex interplay between roots and soil factors such as water, oxygen and nutrients, the distribution of RLD is an important indicator of the potential water uptake. A parallel between root density distribution and water distribution and uptake under drip irrigation is illustrated in Fig. 5. Sonneveld and Voogt (1990) reported that in split-root systems of tomato plants, different proportions of water and ions were taken up by roots from locations of low and medium ion concentrations. Higher proportions of water and ions were taken up from locations in soil with lower salinity. Similar results were obtained for cucumber, although the critical electrical conductivity that reduced water uptake was lower in cucumber than tomato (Sonneveld and De Kreij, 1999). A review of numerous studies that used split-root systems with different salinity concentrations indicated that the water uptake by roots from the least saline part of the soil was the factor driving shoot growth (Bazihizina et al., 2012).

Drip irrigation supplies plants from point or line sources that often overlap and interact. The design of drip systems and understanding the dynamics and patterns of flow and transport from them require the prediction of water movement and solute transport from interacting sources (Merrill et al., 1978; Mmoloawa and Or, 2000). Modeling approaches to predict flow and transport dynamics from drip irrigation have been reviewed by Mmoloawa and Or (2000), Lubana and Narda (2001), Cook et al. (2006) and Subbiah (2013).

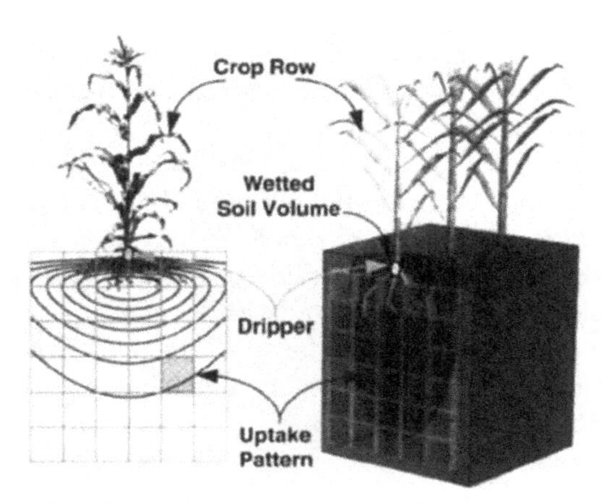

Figure 5 A scheme of uptake intensity patterns (contour lines) and discrete sink terms (grid cells) based on water uptake functions for a cross section of wetted soil perpendicular to drip line direction. Source: adapted from Coelho and Or (1997).

They show a range of empirical, analytical, semi-analytical and numerical methods. Actual water, fertilizer and salt distribution in the soil around a dripper depends on soil hydraulic properties, dripper characteristics, initial conditions, application frequency, evaporation and root uptake. In the case of subsurface drip, inlet pressure and dripper outlet-soil interface geometry and size can also be factors. The desired wetting patterns can be obtained by manipulating dripper flow rate and spacing (Lubana and Narda, 2001) and by influencing the soil dripper interface (Ben-Gal et al., 2004; Meshkat et. al., 2000). Mathematical models can be used to increase the capacity to predict water, fertilizer and salt movement and distribution in soil (Mmoloawa and Or, 2000; Subbiah, 2013). In practice, it is quite a difficult task to represent accurately the temporal changes in solution concentrations in bulk and rhizosphere soils. Roots grow and are active where the soil environment is optimal regarding the availability of water, nutrients and oxygen and with low salinity. Uptake, along with the flow and transport from sources, alters the root-zone environment, including the moisture status and concentration of solutes. Consequently, understanding and simulating transient root water and solute uptake is still a challenge. Obviously, the entire system with irrigation emitters, the soil and its properties, water quality and root distribution and function, including feedback to its environment, must be considered together over time to predict successfully or understand fertigation water and solute dynamics.

5 The right source: fertilizer type and source

The rate of the dissolution of solid fertilizers, as well as their total solubility, influences their use in fertigation. The fertilizer used in fertigation has to be chosen carefully. Compatibility with the ion composition of the irrigating water must be considered. An improper choice may cause difficulties by clogging or corroding the irrigation system, resulting in losses of nutrient elements.

When choosing nitrogen fertilizers, the desired ammonium-to-nitrate ratio has to be considered prior to the selection of sources for each. The most concentrated source of nitrogen is anhydrous ammonia or aqua ammonia; however, it may cause clogging of pipes because of increased pH and lowered solubility of calcium and magnesium salts. Ammonia volatilization losses increase with increasing salt concentration. Thus, anhydrous ammonia can be considered just for high-quality water containing low salt (characterized by solution electrical conductivity of about 0.2 dS/m) and low Ca + Mg concentration (below 10 mg/L). In most cases the sources of N are urea, ammonium nitrate, ammonium sulfate and multi-nutrient fertilizers such as potassium nitrate, calcium nitrate and ammonium phosphate. Multi-nutrient fertilizers are preferred when salt concentration in water is high.

When choosing P fertilizers for fertigation, care must be taken to avoid P-Ca and P-Mg precipitation in pipes and emitters. Therefore, acidic P fertilizers (e.g. phosphoric acid or mono-ammonium phosphate) are recommended because the monovalent ($H_2PO_4^-$) Ca and Mg salts are far more soluble than the divalent (HPO_4^{2-}) ones.

Potassium application in irrigation water is relatively problem free due to the high solubility of most potassium salts. Potassium chloride solubility is higher than that of potassium sulfate, and in water with high calcium concentration, there is a possibility of calcium sulfate precipitation. Conversely, when chloride concentration in water is already high, potassium chloride is not recommended for chloride-sensitive crops. Potassium nitrate is a very efficient source of both K and N.

Nutrients are introduced into irrigation water by injecting concentrated fertilizers, usually regulated by an irrigation computer/controller. Maintaining solubility in the concentrated solutions can be challenging where the solubility of a mixture of fertilizers is usually lower than that of the single fertilizer components. One method for fertilization with multiple kinds of soluble fertilizers is the use of separate tanks for nutrient sources. This method avoids mixing incompatible fertilizers that may lead to the precipitation of poorly soluble salts in a fertilizer tank. The use of multiple tanks is common in greenhouses but is usually too expensive for open-field crops. Typically, at least two tanks that separate phosphorus fertilizers from base cations are used to eliminate the precipitation of metal-phosphorus salts. Tables of reduced solubility and incompatibility of fertilizers are available to aid in the preparation of stock solutions (Matson and Lieth, 2019). Recipes for different crops in different environments are also available (Matson and Lieth, 2019).

An alternative approach is the use of ready-made liquid mixtures of fertilizers provided by fertilizer manufacturers. Tailor-made mixtures for specific requirements of growers are becoming popular in the fertigation of high-value intensive cropping systems. An additional aspect to be considered is a decrease in the solubility of fertilizers as a function of decreasing temperature, dictating the need to adjust fertilizer concentrations to environmental temperature.

Complete nutrient solutions containing micronutrients as well as macronutrients are very common in the intensive production of horticultural crops grown in greenhouses. To avoid precipitation at $pH > 5$ and to facilitate sufficient transport toward roots in the soil, micronutrients are added in solutions as chelates of organic ligands. The main chelating agents used in fertigation systems are Ethylenediaminetetraacetic acid (EDTA, $C_{10}H_{16}O_8N_2$), diethylenetriaminepentaacetic acid (DTPA, $C_{14}H_{23}O_{10}N_3$) and ethylenediaminedi-*o*-hydroxyphenylacetic acid (EDDHA, $C_{18}H_{20}O_6N_2$). The synthetic chelates lose their capacity to retain microelements at $pH < 3$. This poses a problem in stock solutions; therefore, it is advisable to keep chelated microelements separately from acid fertilizers at pH > 5 and inject them into the water after the acid stock solutions.

5.1 Nitrogen sources

There are two ionic forms of N taken up by plants: NH_4^+ and NO_3^-. The form-dependent potential problems of N sources are ammonia toxicity, modification of the rhizosphere pH, changes in the availability of other nutritional elements and incidence of physiological disorders such as chlorosis and blossom-end rot. Ammonium toxicity has been hypothesized to be associated with the high solution concentration of ammonia (NH_3(*aq*)) (Bennett and Adams, 1970a,b). NH_3(*aq*) can enter plants across the root-cell plasma membranes easily and, even at a relatively low concentration (0.17 mM in the nutrient medium), might inhibit vital processes such as photosynthesis, respiration and enzyme activity (Bennett and Adams, 1970b). The inhibition of root elongation as a result of NH_4^+ nutrition was found to be pH dependent and directly related to the calculated solution concentrations of NH_3(*aq*) (Bennett and Adams, 1970b).

Nitrogen sources may significantly affect rhizosphere pH (Nye, 1981; Bar-Yosef, 1999; Bloom et al., 2003), especially in the restricted wet volume under drip irrigation (Bar-Yosef, 1999). The common N fertilizer sources in soil systems are urea, NH_4^+ and NO_3^-. Nitrogen sources may affect the rhizosphere pH via three mechanisms (Bar-Yosef, 1999): (i) displacement of H^+/OH^- adsorbed on the solid phase, (ii) nitrification/denitrification reactions and (iii) release or uptake of H^+ by roots in response to NH_4 or NO_3 uptake. Mechanisms (i) and (ii) are not associated with plant activity and affect the whole volume of the fertigated substrate, whereas mechanism (iii) is directly related to the uptake of nutritional elements and may be very effective because it affects a limited volume in the immediate vicinity of the roots (Bloom et al., 2003). The extent of the pH alterations caused by the three mechanisms described earlier depends on soil properties, wetted volume, plant activity and the environmental factors that affect the nitrification rate.

Ammonium uptake is preferred over nitrate by most plants (Hawkesford et al., 2012). However, in the soil, the nitrification rate is rapid and therefore the NH_4^+ concentration diminishes quickly. Frequent fertigation by drip irrigation maintains NO_3/NH_4 ratio (R_N) in the soil similar to that in the irrigation water. Consequently, under high-frequency fertigation, the impact of varying NH_4^+ concentration in irrigation solutions and the R_N on rhizosphere pH and physiological processes becomes crucial; hence, controlling the R_N is of utmost importance. The reduction of the soil and rhizosphere pH, driven by NH_4^+ nitrification and root excretion of protons, can be used as a tool for overcoming growth disorders induced by micronutrient deficiencies, such as chlorosis and 'little leaf' or 'rosette' (Savvas et al., 2003; Silber et al., 1998, 2003).

Nitrogen sources have a significant direct effect on numerous physiological processes in plants (Hawkesford et al., 2012). It has been widely reported that NH_4^+ nutrition depressed the uptake of cations, especially in leaves and petioles, whereas NO_3^- nutrition depressed the uptake of anions. As nitrogen is the major macronutrient, the cation-anion balance in plant tissues is controlled by the ratio of ammonium-to-nitrate uptake (Hawkesford et al., 2012). The R_N affects the apoplastic pH and, consequently, Fe^{III} reduction and mobilization in plants (Kosegarten et al., 2004). Mengel (1994) found that the inhibition of iron uptake from the root apoplast into the cytosol of root cells, rather than low iron availability in the rhizosphere, was the main cause of iron chlorosis. High NO_3^- in the rhizosphere induced chlorosis even when the concentration of iron in plants was normal and not different from that in non-chlorotic plants irrigated with high R_N solution (Kosegarten et al., 2004).

6 Models and decision support tools for design, operation and optimization of fertigation systems

There are a plethora of models simulating water, salinity and nutrient (mainly N) dynamics, including for fertigation systems. These have been shown to successfully simulate cropping systems and, as such, have been suggested as decision support tools for fertigation management (Bar-Yosef et al., 2004, 2006; Phogat et al., 2014; Gallardo et al., 2009; Rahn et al., 2010; Conversa et al., 2015; Elia and Conversa, 2015; Giménez et al., 2013). Due to their complexity and the need to develop and evaluate them under real-time conditions (Bar-Yosef et al., 2004), such models have been used primarily as planning support in the evaluation and comparison of different scenarios and less as the day-to-day decision-making tools.

Complex soil hydraulic and/or plant physiological models for the analysis of the fertigation management alternatives are sensitive to input parameters and basic model assumptions. For example, the question of optimizing fertigation scheduling was addressed by two separate research teams, both of whom evaluated solute distribution from drip irrigation (Cote et al., 2003; Gärdenäs et al., 2005). Both asked the question: During an irrigation event, when would application of fertilizer be most efficient, leading to the greatest amount of uptake by crops and least leaching beyond the root zone? Both teams used the identical means to evaluate the management strategies: the HYDRUS-2D (Simunek et al., 1999) model as a predominant, accessible and strong tool for the simulation of vadose zone water and solute dynamics. Interestingly, Cote et al. (2003) concluded that fertigation at the beginning of the irrigation cycle reduces leaching, whereas Gärdenäs et al. (2005) found the opposite, that fertigation at the beginning of irrigation cycles causes more leaching than at the end. Even though the major difference between the studies involved assumptions and parameterization regarding root placement and uptake, other variables in the conditions defined for the simulations are also liable to influence results. The use of models such as HYDRUS-2D for the evaluation of fertigation strategies therefore must be evaluated critically on a case-by-case basis.

Models to predict specific response to environmental variables including fertigation, water quality and irrigation scheduling have been developed for specific crops in specific growing systems and have been suggested as possible decision support systems (DSSs). Following are a few examples. Bar-Yosef et al. (2004, 2006) simulated tomato production in greenhouse with closed system fertigation. Voogt et al. (2006) presented a 'Fertigation Model' as a DSS for water and nutrient supply for soil-grown greenhouse crops. The TOMGRO model (Gallardo et al., 2009) simulated N uptake by tomatoes in greenhouses. An additional DSS for fertigation management in open-field vegetable crops, GesCoN (Elia and Conversa, 2015; Conversa et al., 2015), has been promoted more recently. Barradas et al. (2012) built DSS-FS (decision support system-fertigation simulator) to aid in designing drip and sprinkler systems, compute reference ET and estimate nutrition requirements based on soil fertility data and crop uptake tables. Not specific to fertigation, the EU-Rotate_N DSS (Rahn et al., 2010) is interesting as it has an extensive collection of crop modules and incorporates root growth simulation and N chemistry as it assesses economic and environmental performance of crop rotations for more than 70 arable and horticultural crops and a wide range of growing systems and conditions. Each of these models and their decision support tools, and others like them, could be used to promote improved nutrient (mainly N) management, provided they are properly parameterized and adapted/developed for the specific crop and cropping system in question (Battilani and Fereres, 1998).

Lately, simulation models are simplified and converted to online or mobile fertigation applications, making them ever more accessible to end users. A noticeable example is the VegSyst (Giménez et al., 2013; Gallardo et al., 2014) model, first introduced as a crop growth, N uptake and ET model, adapted as an on-farm DSS for pepper and tomato crops. VegSyst was recently expanded in a large European Union-funded effort (FERTINNOWA https://www.fertinnowa.com/) to a number of greenhouse crops including tomato (*S. lycopersicum*), pepper (*C. annuum*), cucumber (*Cucumis sativus*), zucchini (*Cucurbita pepo*), trellised and non-trellised melon (*Cucumis melo* L.), watermelon (*Citrullus lanatus*) and eggplant (*Solanum melongena* L.) and is applicable with soil- or substrate-grown crops. The software has only few inputs and is simple to use. Calculations can be done using long-term average climate data, and the plans of N and irrigation applications for the duration of the crop can be made prior to planting. Input data are meteorological data (air temperature and solar radiation), crop and cropping dates, and details of soil, irrigation system and most recent manure application. The output is the daily amount of irrigation and N to apply and the average N concentration to be applied by fertigation during 4-week periods. The authors suggest that such prescriptive management should be supplemented with simple soil/plant monitoring techniques to enable fine-tuning of irrigation and N management.

7 Case studies

The previous sections presented the relevant background and principles of fertigation. In this section we present three case studies that demonstrate sustainable and efficient fertigation systems. The first and second case studies present how the manipulation of the N source is an effective means for reducing the pH of the soil in orchard and soilless culture, leading to improved growth conditions and the enhancement of the uptake of nutrients whose availability tends to be low. The third case study demonstrates the capacity to save water and nutrients in a fertigated recycling system in comparison to conventionally grown plants in the soil. Each of the case studies demonstrates unique features of the fertigation method.

7.1 Grapefruit orchard

The effects of N source and nitrification inhibitor (NI) on soil and trees were investigated in a commercial orchard of mature grapefruit (*Citrus × paradisi*) trees planted on a sandy loam soil in the center of Israel. The experiment was conducted during 3 years, each including three irrigation seasons (May to October) and two rainfall seasons (November to April). The trees were spaced at 6 × 4 m (between and in rows) and drip-irrigated with one lateral pipe per tree row, 0.5 m between adjacent drippers and a discharge rate of 3.8 L/h. There were three treatments: fertigation with ammonium nitrate (AN), ammonium sulfate (AS) or ammonium sulfate with NI (AS + NI). All fertilizers and NI were applied through the irrigation system. The N inhibitors were DCD at 10% w/w of the N rate during the first 2 years, and DMPP at 1% w/w of the N rate during the 3rd year. Each treatment was replicated five times in a complete randomized block design. For more details see Erner et al. (2017).

Fertigation with AS and AS + NI resulted in a reduction of pH in the soil solution in the grapefruit orchard in the irrigation season (from 7.5 to 5.5–6.5), which was accompanied by increased concentrations of P and Mn in the soil solution (Fig. 6). During the winter (rainfall season) the soil solution pH in the

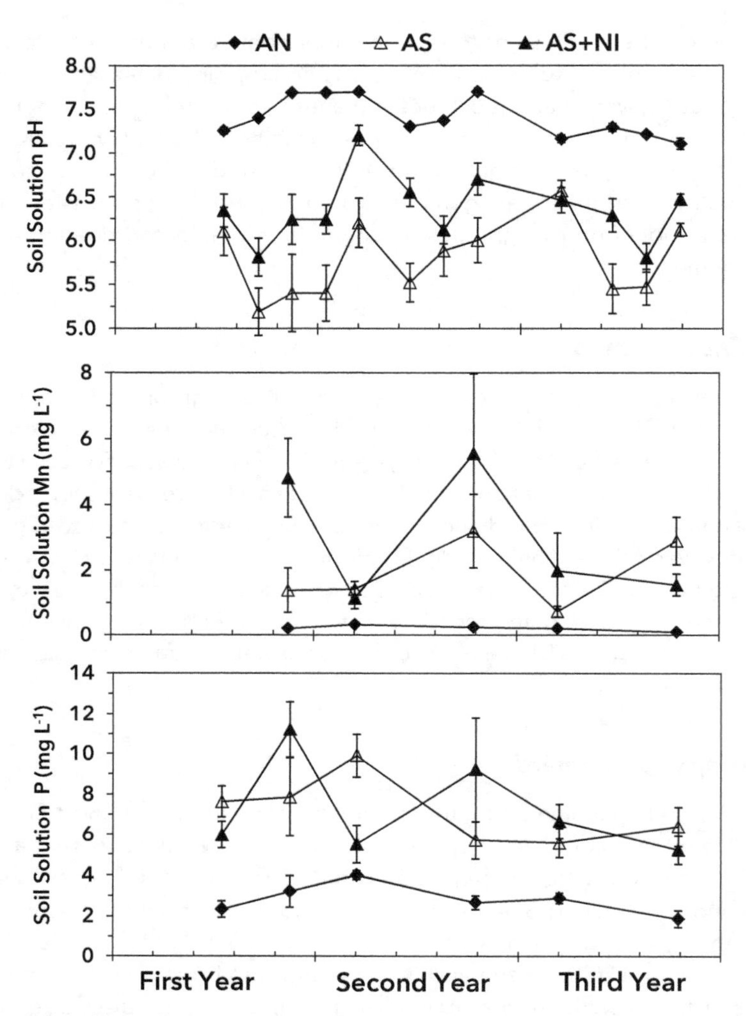

Figure 6 The effects of nitrogen source (AN - ammonium nitrate, AS - ammonium sulfate, AS+NI - ammonium sulfate + nitrification inhibitor) in the irrigation water of a grapefruit orchard on soil solution pH and P and Mn concentration. Source: adapted from Erner et al. (2017).

AS and AS + NI treatment increased almost back to the initial values. The tissue concentrations of P and Mn in grapefruit leaves were significantly higher in the AS and AS + NI treatment than the AN treatment (Erner et al., 2017). In addition, under AS and AS + NI treatment, P and Mn uptake, chlorophyll content in leaves and the skin of green fruit, and total soluble solids (TSSs) in fruits were higher than those in the AN treatment (Erner et al., 2017). The application of NI enhanced the capacity of AS to raise N uptake above that achieved with AN and reduced the potential for N leaching from the soil, enhancing the sustainability of grapefruit production.

7.2 Chives in soilless culture

The effects of N source and the pH of irrigation solution on the mineral uptake are illustrated in a system where chive (*Allium schoenoprasum*) was grown on a perlite medium and irrigated with solutions containing different fractions of ammonium nitrogen and in one treatment acidified by sulfuric acid. The overall concentration of nitrogen supplied to all treatments was 100 mg N/L.

Table 1 presents the effects of increased NH_4-N in the irrigation of chives from 26% to 69%, and the effect of acidifying the solution of low percentage of NH_4-N (29%) from 7.1 to 4.9 by sulfuric acid. The reduction in NH_4-N concentration in the drainage relative to its concentration in the irrigation water indicates that the uptake of NH_4 by the chives was faster than that of NO_3. We assume that the uptake of NH_4 decreased as the pH in the irrigation solution was acidified from 7.1 to 4.9. The expected effect of the percentage of NH_4-N on the pH was observed, the drainage pH decreased when the percentage of NH_4-N of the irrigating solution was 43 and 69 and it increased when it was 26. Direct acidification by sulfuric acid significantly increased the concentrations of P, Fe, B and Cu in the leaves. Acidification by the high percentage of NH_4-N increased significantly the concentrations of P and B in chive leaves. This can be explained by the effect of NH_4-N on the pH, in agreement with the results of the first case study and previous publications (Erner et al., 2017; Nye, 1981; Bar-Yosef, 1999; Bloom et al., 2003). The concentrations of Fe and Cu were not affected significantly by the percentage of NH_4-N, probably due to the high variation in their concentrations. As expected, N concentration was not influenced by neither R_N nor acidification. It remained unclear as to why K and Mn concentrations in leaves were not affected by the pH and the percentage of NH_4-N in the irrigation solution.

7.3 Chives in a recycled irrigation system

Growing crops in a system where water draining out of the root zone is collected and re-used for irrigation (recycling system) could potentially result in water- and nutrient-use efficiency. This is the main method used in the Netherlands where a policy of zero nutrient and salt loads from greenhouses is enforced (Voogt et al., 2013). The composition of the irrigating water in the recycled system is not constant over time due to the differences in the

Table 1 Mineral concentration in chive (*A. schoenoprasum*) leaves grown on a perlite medium and irrigated with solutions containing different percentages of N-NH_4 and acidified by sulfuric acid (the first treatment)

Irrigation			Drainage			Mineral concentration						
N-NH_4	N-NO_3		N-NH_4	N-NO_3		N	P	K	Fe	Mn	B	Cu
%		pH	%		pH	mg/g DW				μg/g DW		
29	71	4.9[a]	18	82	5.4	40.4	6.5 A	43.6	212 A	79.3	6.5 A	59.0 A
26	74	7.1	0	100	7.7	39.6	5.5 B	42.9	118 B	91.0	5.5 B	31.5 B
43	57	7.1	6	94	6.5	39.7	5.9 AB	42.2	128 AB	87.0	5.9 AB	36.5 B
69	31	7.1	32	82	5.1	41.3	6.8 A	42.0	173 AB	92.2	6.8 A	35.6 B

[a] Acidification of the irrigating solution by sulfuric acid.
In all four treatments, the total N supplied was 100 mg N/L.
Each value is an average of five replicates (in blocks). Treatments accompanied with the same letter are not significantly different according to Tukey's honestly significant difference two-way analysis of variance with JMP 10.0 software (SAS Institute Inc., Cary, NC). Default significance levels were set at a value of 0.05.
n.s. = non-significant.
Source: adapted from Yermiyahu et al. (2005).

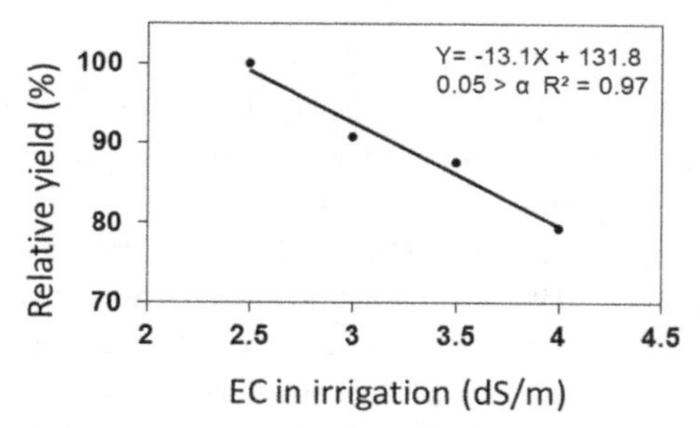

Figure 7 Relative yield of chive (*A. schoenoprasum*) in a recycling system as affected by the electrical conductivity of irrigation water.

rates of the uptake of water and the different minerals. In order to have an efficient system, optimal range of water properties including pH, salinity and mineral concentrations must be defined. One known phenomenon in recycling systems is an increase in the concentration of total ions (salinity) with time in the recycled solution. This was illustrated in a recycling system where chive was grown on a perlite medium. Increasing salinity (electrical conductivity, EC) decreased yield at a rate of 13% for every unit increase in EC (dS/m) (Fig. 7).

It was found that growing chives in a recycling system can save up to 70% of water compared to recommended water head for growing chives in a non-recycled system (in soil) during the same season. Growing chives in the recycling system was also found to save a significant amount of fertilizers (Table 2). For example, irrigation with 100 mg N/L (common farmer practice) in a recycling system saved 45% of N compared to growth in the soil. It was also found that growing in the recycling system allowed a decrease of N in the irrigation solution by 75% to 25 mg N/L. At this N concentration N saving was enhanced to 62%. Similar results were found for P (Table 2).

Table 2 Nitrogen and phosphorus savings for chives (*A. schoenoprasum*) grown in a recycling system relative to a non-recycled system (in the soil)

Mineral in irrigation (mg/L)		Mineral saving (%)	
Nitrogen	Phosphorus	Nitrogen	Phosphorus
100	15	45	-
25	15	62	-
100	15	-	42
100	5	-	69

Mineral savings were calculated based on the total mineral use in the recycling system compared to recommended fertilization in non-recycled system (in soil).

8 Conclusion

The main advantages of fertigation over irrigation, combined with broadcast or banding fertilizers for sustainable fertilization, can be summarized as follows: (a) the application of nutrients and water is accurate and uniform under all circumstances; (b) application is restricted to the wetted area, where the root activity is concentrated; (c) the amounts and concentrations of specific nutrients can be adjusted to crop requirements according to the stage of development and climatic conditions; (d) fluctuations in nutrient concentrations in the soil over the course of the growing season are reduced; (e) crop foliage is kept dry, thus restricting the development of plant pathogens and reducing the use of pesticides; (g) energy use is decreased by the avoidance of broadcast operations (although a life cycle analysis is required for the analysis of the balance of energy saving with the additional input of energy used in producing, transporting, installing and maintaining the fertigation systems); (h) soil compaction and mechanical damage to crops are reduced because there is less tractor traffic; and (i) convenient use is enabled of the compound, ready-mixed and balanced liquid fertilizers, with low concentrations of micronutrients that are otherwise difficult to apply accurately in the field.

In presenting the principles of fertigation and the case studies, we showed how fertigation contributes to sustainable agriculture by maximizing the outcome from each unit of fertilizer and from each drop of water. However, fertigation is not a miracle solution and cannot overcome or compensate for all the negative effects of modern and industrial agriculture on sustainability, including a decline in soil organic matter, salinization and environmental pollution by pesticides (Celik, 2005; Lal, 2002, 2004). We suggest that fertigation should be combined with various additional proven means for promoting sustainable agriculture: leaving and incorporating crop residues in the field, amending soil with organic wastes and composts, no or minimum tillage, crop rotation, cover crops, minimizing pesticide application (Drinkwater et al., 1998; Lal, 2002, 2004; Halpern et al., 2010; Madgoff and Weil, 2004), and cultivation of perennial crops that provide organic matter into the soil by root exudates and degradation (Crews and Rumsey, 2017).

9 Future trends

Staple food production with fertigation – there is ongoing increase in demand for staple food production due to the population growth and the rise in the standard of living. To meet this demand some of the rain-fed areas, where new sources of water (like treated waste water and desalinated water) are available, will be transformed to irrigated cultivation. and surface flooding (rice) and furrow (maize) will be replaced by mini-sprinklers and drip irrigation, where fertigation is the most efficient and sustainable practice. Research is required to understand nutritional needs and crop responses to fertigation rates.

Fertigation with low-pressure systems. Constrains on power production and distribution in developing countries and remote areas limit the use of fertigation. Development and production of low-pressure systems will enable the implementation of fertigation in these areas.

Precision agriculture. In the current methods of fertigation, the supply of water and nutrients aims to meet the average requirements of plants in an irrigated field. Hence, conditions of deficiency and surplus may coexist in different parts of the same field. Future research should be directed to meeting the spatial variability in water and nutrient requirements of crops in a given field.

DSSs based on models and using remote-sensing and weather information. Fertigation provides the means to supply the right fertilizers at the right rate, time and place, but reliable DSSs are required to guide best-practice management decisions by farmers for the optimal use of fertigation in the field.

Sensors for mineral nutrition status of crops and fertigation needs. Progress in the proximal and remote-sensing methods, combined with reliable DSSs, will enable more exact and efficient fertigated supply of nutrients at the right rate, time and place.

Fertigation with nanofertilizers. One of the pillars of fertigation is the right source of fertilizers, with their solubility being a critical factor. Future advances in nanotechnology offer a promising way to produce new fertilizers effective in fertigation.

10 References

Alexandratos, N. and Bruinsma, J. 2012. World agriculture towards 2030/2050: the 2012 revision. ESA Working Paper No. 12-03. FAO, Rome.

Assouline, S. 2002. The effects of microdrip and conventional drip irrigation on water distribution and uptake. *Soil Sci. Soc. Am. J.* 66(5), 1630–6. doi:10.2136/sssaj2002.1630.

Assouline, S., Möller, M., Cohen, S., Ben-Hur, M., Grava, A., Narkis, K. and Silber, A. 2006. Soil-plant system response to pulsed drip irrigation and salinity: bell pepper case study. *Soil Sci. Soc. Am. J.* 70(5), 1556–68. doi:10.2136/sssaj2005.0365.

Barradas, J. M., Matula, S. and Dolezal, F. 2012. A decision support system-fertigation simulator (DSS-FS) for design and optimization of sprinkler and drip irrigation systems. *Comput. Electron. Agric.* 86, 111–9. doi:10.1016/j.compag.2012.02.015.

Bar-Tal, A., Feigin, A., Rylski, I. and Pressman, E. 1994. Effects of root pruning and N-NO_3 solution concentration on nutrient uptake and transpiration of tomato plants. *Sci. Hortic.* 58(1–2), 77–90. doi:10.1016/0304-4238(94)90129-5.

Bar-Tal, A., Aloni, B., Karni, L. and Rosenberg, R. 2001. Nitrogen nutrition of greenhouse pepper: II. Effects of nitrogen concentration and NO_3:NH_4 ratio on growth, transpiration, and nutrient uptake. *HortScience* 36(7), 1252–9. doi:10.21273/HORTSCI.36.7.1252.

Bar-Tal, A., Yermiyahu, U., Beraud, J., Keinan, M., Rosenberg, R., Zohar, D., Rosen, V. and Fine, P. 2004. N, P and K uptake by wheat and their distribution in soil following successive annual application of composts. *J. Environ. Qual.* 33, 1855–65.

Bar-Yosef, B. 1999. Advances in fertigation. *Adv. Agron.* 65, 1–77. doi:10.1016/S0065-2113(08)60910-4.

Bar-Yosef, B. and Sheikholslami, M. R. 1976. Distribution of water and ions in soils irrigated and fertilized from a trickle source. *Soil Sci. Soc. Am. J.* 40(4), 575–82. doi:10.2136/sssaj1976.03615995004000040033x.

Bar-Yosef, B., Fishman, S. and Kläring, H. P. 2004. A model-based decision support system for closed irrigation loop greenhouses. *Acta Hortic.* 654, 107–22. doi:10.17660/ActaHortic.2004.654.11.

Bar-Yosef, B., Fishman, S. and Kläring, H. P. 2006. A model describing root growth and water, N and Cl uptake in closed loop irrigation systems. *Acta Hortic.* 718, 435–44. doi:10.17660/ActaHortic.2006.718.50.

Battilani, A. and Fereres, E. 1998. The use of decision support systems to manage fertigation and to minimize environmental effects: a challenge for the future. *Acta Hortic.* 487, 547–56.

Bazihizina, N., Barrett-Lennard, E. G. and Colmer, T. D. 2012. Plant growth and physiology under heterogeneous salinity. *Plant Soil* 354, 1–19.

Ben-Gal, A. and Dudley, L. M. 2003. Phosphorus availability under continuous point source irrigation. *Soil Sci. Soc. Am. J.* 67(5), 1449–56. doi:10.2136/sssaj2003.1449.

Ben-Gal, A. and Lazarovitch, N. 2003. Beyond burying the lateral: current issues in and future opportunities for subsurface drip irrigation. Prepared for Netafim Irrigation Company.

Ben-Gal, A., Lazarovitch, N. and Shani, U. 2004. Subsurface drip irrigation in gravel filled cavities. *Vadose Zone J.* 3(4), 1407–13. doi:10.2113/3.4.1407.

Bennett, A. C. and Adams, F. 1970a. Calcium deficiency and ammonia toxicity as separate causal factors of $(NH_4)_2HPO_4$ injury to seedlings. *Soil Sci. Soc. Am. Proc.* 34(2), 255–9. doi:10.2136/sssaj1970.03615995003400020023x.

Bennett, A. C. and Adams, F. 1970b. Concentration of NH_3(aq) required for incipient NH_3 toxicity to seedlings. *Soil Sci. Soc. Am. Proc.* 34(2), 259–63. doi:10.2136/sssaj1970.03615995003400020024x.

Beraud, J., Fine, P., Keinan, M., Rosenberg, R., Hadas, A. and Bar-Tal, A. 2005. Modeling carbon and nitrogen transformations for adjusting compost application with N consumption by wheat. *J. Environ. Qual.* 34, 664–75.

Bloom, A. J., Meyerhoff, P. A., Taylor, A. R. and Rost, T. L. 2003. Root development and adsorption of ammonium and nitrate from the rhizosphere. *J. Plant Growth Regul.* 21, 416–31.

Bravdo, B. A., Levin, I. and Assaf, R. 1992. Control of root size and root environment of fruit trees for optimal fruit production. *J. Plant Nutr.* 15(6–7), 699–712. doi:10.1080/01904169209364356.

Bresler, E. 1978. Analysis of trickle irrigation with application to design problems. *Irrig. Sci.* 1(1), 3–17. doi:10.1007/BF00269003.

Celik, I. 2005. Land-use effects on organic matter and physical properties of soil in southern Mediterranean highland of Turkey. *Soil Till. Res.* 83(2), 270–7. doi:10.1016/j.still.2004.08.001.

Chandra, S. P. and Rai, A. K. 1996. Nonlinear root-water uptake model. *J. Irrig. Drain. Engg. ASCE* 122(4), 198–202. doi:10.1061/(ASCE)0733-9437(1996)122:4(198).

Clothier, B. E. and Green, S. R. 1994. Rootzone processes and the efficient use of irrigation water. *Agric. Water Manag.* 25(1), 1-12. doi:10.1016/0378-3774(94)90048-5.

Coelho, F. E. and Or, D. 1997. Applicability of analytical solutions for flow from point sources to drip irrigation management. *Soil Sci. Soc. Am. J.* 61(5), 1331-41. doi:10.2136/sssaj1997.03615995006100050007x.

Conversa, G., Bonasia, A., Di Gioia, F. and Elia, A. 2015. A decision support system (GesCoN) for managing fertigation in open field vegetable crops. Part II–model calibration and validation under different environmental growing conditions on field grown tomato. *Front. Plant Sci.* 6, 495. doi:10.3389/fpls.2015.00495.

Cook, F. J., Fitch, P., Thorburn, P. J., Charlesworth, P. B. and Bristow, K. L. 2006. Modelling trickle irrigation: comparison of analytical and numerical models for estimation of wetting front position with time. *Environ. Modell. Softw.* 21(9), 1353-9. doi:10.1016/j.envsoft.2005.04.018.

Cote, C. M., Bristow, K. L., Charlesworth, P. B., Cook, F. J. and Thorburn, P. J. 2003. Analysis of soil wetting and solute transport in subsurface trickle irrigation. *Irrig. Sci.* 22(3-4), 143-56. doi:10.1007/s00271-003-0080-8.

Crews, T. E. and Rumsey, B. 2017. What agriculture can learn from native ecosystems in building soil organic matter: a review. *Sustainability* 9(4), 578. doi:10.3390/su9040578.

Dahan, O., Babad, A., Lazarovitch, N., Russak, E. E. and Kurtzman, D. 2014. Nitrate leaching from intensive organic farms to groundwater. *Hydrol. Earth Syst. Sci.* 18(1), 333-41. doi:10.5194/hess-18-333-2014.

Drinkwater, L. E., Wagoner, P. and Sarrantonio, M. 1998. Legume-based cropping systems have reduced carbon and nitrogen losses. *Nature* 396(6708), 262-5. doi:10.1038/24376.

El-Hawary, A. 2005. Best management practices for the drainage water reuse. A Technical Report. International Center for Agricultural Research in the Dry Areas, 57pp.

Elia, A. and Conversa, G. 2015. A decision support system (GesCoN) for managing fertigation in open field vegetable crops. Part I-methodological approach and description of the software. *Front. Plant Sci.* 6, 319. doi:10.3389/fpls.2015.00319.

Elliott, S. 2016. The nutrient challenge of sustainable fertilizer management. United States Department of Agriculture, National Institute of Food and Agriculture. Available at: https://nifa.usda.gov/blog/nutrient-challenge-sustainable-fertilizer-management (accessed on 17 November 2019).

Erner, Y., Bar-Tal, A., Tagari, E., Levkowich, I. and Bar-Yosef, B. 2017. Nitrogen source and nitrification inhibitors affect soil nutrient status and 'star ruby' grapefruit performance. *Citrus Research and Technology* 37(2), 182-93.

FAO. 2017. *The Future of Food and Agriculture - Trends and Challenges*. Food and Agriculture Organization of the United Nations, Rome.

Gallardo, M., Thompson, R. B., Rodríguez, J. S., Rodríguez, F., Fernández, M. D., Sánchez, J. A. and Magán, J. J. 2009. Simulation of transpiration, drainage, N uptake, nitrate leaching, and N uptake concentration in tomato grown in open substrate. *Agric. Water Manag.* 96(12), 1773-84. doi:10.1016/j.agwat.2009.07.013.

Gallardo, M., Thompson, R. B., Giménez, C., Padilla, F. M. and Stöckle, C. O. 2014. Prototype decision support system based on the VegSyst simulation model to calculate crop N and water requirements for tomato under plastic cover. *Irrig. Sci.* 32(3), 237-53. doi:10.1007/s00271-014-0427-3.

Gärdenäs, A. I., Hopmans, J. W., Hanson, B. R. and Šimůnek, J. 2005. Two-dimensional modeling of nitrate leaching for various fertigation scenarios under micro-irrigation. *Water Manag.* 74(3), 219–42. doi:10.1016/j.agwat.2004.11.011.

Giménez, C., Gallardo, M., Martínez-Gaitán, C., Stöckle, C. O., Thompson, R. B. and Granados, M. R. 2013. VegSyst, a simulation model of daily crop growth, nitrogen uptake and evapotranspiration for pepper crops for use in an on-farm decision support system. *Irrig. Sci.* 31(3), 465–77. doi:10.1007/s00271-011-0312-2.

Halpern, M. T., Whalen, J. K. and Madramootoo, C. A. 2010. Long-term tillage and residue management influences soil carbon and nitrogen dynamics. *Soil Sci. Soc. Am. J.* 74(4), 1211–7. doi:10.2136/sssaj2009.0406.

Hassan, G. M., Reneau, R. B. and Hagedorn, C. 2010. Solute transport dynamics where highly treated effluent is applied to soil at varying rates and dosing frequencies. *Soil Sci.* 175(6), 278–92. doi:10.1097/SS.0b013e3181e73be8.

Hawkesford, M., Horst, W., Kichey, T., Lambers, H., Schjoerring, J., Skrumsager Møller, I. and White, P. 2012. Functions of macronutrients. In: Marschner, P. (Ed.), *Marschner's Mineral Nutrition of Higher Plants*. Elsevier Academic Press, London, UK, Waltham, MA and San Diego, CA, pp. 135–89. Chapter 6.

Hayhoe, H. 1981. Analysis of a diffusion model for plant root growth and an application to plant soil-water uptake. *Soil Sci.* 131(6), 334–43.

Hochmuth, G. J. 1992. Fertilizer management for drip-irrigated vegetables in Florida. *HortTechnology* 2(1), 27–32. doi:10.21273/HORTTECH.2.1.27.

Kang, Y., Wang, F., Liu, H. and Yuan, B. 2004. Potato evapotranspiration and yield under different drip irrigation regimes. *Irrig. Sci.* 23(3), 133–43. doi:10.1007/s00271-004-0101-2.

Kargbo, D., Skopp, J. and Knudsen, D. 1991. Control of nutrient mixing and uptake by irrigation frequency and relative humidity. *Agron. J.* 83(6), 1023–8. doi:10.2134/agronj1991.00021962008300060018x.

Klepper, B. 1991. Corn root response to irrigation. *Irrig. Sci.* 12, 105–8.

Kosegarten, H., Hoffmann., B., Rroco, E., Grolig, F., Glüsenkamp, K. H. and Mengel, K. 2004. Apoplastic pH and FeIII reduction in young sunflower (*Helianthus annuus*) roots. *Physiol. Plant.* 122(1), 95–106. doi:10.1111/j.1399-3054.2004.00377.x.

Kremmer, S. and Kenig, E. 1996. *Principles of Drip Irrigation*. Irrigation & Field Service. Extension Service, Ministry of Agriculture and Rural Development, Israel (in Hebrew).

Lal, R. 2002. Soil carbon dynamics in cropland and rangeland. *Environ. Pollut.* 116(3), 353–62. doi:10.1016/s0269-7491(01)00211-1.

Lal, R. 2004. Soil carbon sequestration to mitigate climate change. *Geoderma* 123(1–2), 1–22. doi:10.1016/j.geoderma.2004.01.032.

Lamm, F. R. and Trooien, T. P. 2003. Subsurface drip irrigation for corn production: a review of 10 years of research in Kansas. *Irrig. Sci.* 22(3–4), 195–200. doi:10.1007/s00271-003-0085-3.

Li, J., Xu, X., Lin, G., Wang, Y., Liu, Y., Zhang, M., Zhou, J., Wang, Z. and Zhang, Y. 2018. Micro-irrigation improves grain yield and resource use efficiency by co-locating the roots and N-fertilizer distribution of winter wheat in the North China Plain. *Sci. Total Environ.* 643, 367–77. doi:10.1016/j.scitotenv.2018.06.157.

Li, J., Wang, Y., Zhang, M., Liu, Y., Xu, X., Lin, G., Wang, Z., Yang, Y. and Zhang, Y. 2019. Optimized micro-sprinkling irrigation scheduling improves grain yield by increasing the uptake and utilization of water and nitrogen during grain filling in winter wheat. *Agric. Water Manag.* 211, 59–69. doi:10.1016/j.agwat.2018.09.047.

Locascio, S. J., Hochmuth, G. J., Rhoads, F. M., Olson, S. M., Smajstria, A. G. and Hanlon, E. A. 1997. Nitrogen and potassium application scheduling effects on drip-irrigated tomato yield and leaf tissue analysis. *HortScience* 32(2), 230–5. (doi:10.21273/HORTSCI.32.2.230).

Lorenzo, H. and Forde, B. 2001. The nutritional control of root development. *Plant Soil* 232(1/2), 51–68. doi:10.1023/A:1010329902165.

Lubana, P. P. S. and Narda, N. K. 2001. Modelling soil water dynamics under trickle emitters – a review. *J. Agric. Engg. Res.* 78(3), 217–32. doi:10.1006/jaer.2000.0650.

Lynch, J. P., Ho, M. D. and phosphorus, L. 2005. Rhizoeconomics: carbon costs of phosphorus acquisition. *Plant Soil* 269(1–2), 45–56. doi:10.1007/s11104-004-1096-4.

Ma, Z., Bielenberg, D. G., Brown, K. M. and Lynch, J. P. 2001. Regulation of root hair density by phosphorus availability in *Arabidopsis thaliana*. *Plant Cell Environ.* 24(4), 459–67. doi:10.1046/j.1365-3040.2001.00695.x.

Madgoff , F. and Weil, R. 2004. Soil organic matter management strategies. In: Madgoff, F. and Weil, R. (Eds), *Soil Organic Matter in Sustainable Agriculture*. CRC Press, Boca Raton, FL, pp. 45–65.

Matson, N. and Lieth, J. H. 2019. Liquid culture hydroponic system operation. In: Raviv, M., Lieth, J. H. and Bar-Tal, A. (Eds), *Soilless Culture: Theory and Practice*. Elsevier Academic Press, UK, pp. 567–85.

Mengel, K. 1994. Iron availability in plant tissues – iron chlorosis on calcareous soils. *Plant Soil* 165(2), 275–83. doi:10.1007/BF00008070.

Merrill, S. D., Raats, P. A. C. and Dirksen, C. 1978. Laterally confined flow from a point source at the surface of an inhoniogeneous soil column. *Soil Sci. Soc. Am. J.* 42(6), 851–7. doi:10.2136/sssaj1978.03615995004200060002x.

Meshkat, M., Warner, R. C. and Workman, S. R. 2000. Evaporation reduction potential in an undisturbed soil irrigated with surface drip and sand tube irrigation. *Trans. ASAE* 43(1), 79–86. (doi:10.13031/2013.2690).

Mmoloawa, K. and Or, D. 2000. Root zone solute dynamics under drip irrigation: a review. *Plant Soil* 222(1/2), 163–90. doi:10.1023/A:1004756832038.

Molz, F. J. 1981. Models of water transport in the soil-plant system: a review. *Water Resour. Res.* 17(5), 1245–60. doi:10.1029/WR017i005p01245.

Neary, P. E., Storlie, C. A. and Paterson, J. W. 1995. Fertilization requirements for drip-irrigated bell pepper grown on loamy sand soils. In: Lamm, F. R. (Ed.), *Microirrigation for a Changing World: Conserving Resources/Preserving the Environment*. ASAE Publ. 4–95. American Society of Agricultural and Biological Engineers, St. Joseph, MI, pp. 187–93.

Nye, P. H. 1981. Changes of pH across the rhizosphere induced by roots. *Plant Soil* 61, 7–26.

Palacios-Díaz, M. P., Mendoza-Grimón, V., Fernández-Vera, J. R., Rodríguez-Rodríguez, F., Tejedor-Junco, M. T. and Hernández-Moreno, J. M. 2009. Subsurface drip irrigation and reclaimed water quality effects on phosphorus and salinity distribution and forage production. *Agric. Water Manag.* 96, 1659–1666.

Phene, C. J., Davis, K. R., Hutmacher, R. B., Bar-Yosef, B., Meek, D. W. and Misaki, J. 1991. Effect of high-frequency surface and subsurface drip irrigation on root distribution of sweet corn. *Irrig. Sci.* 12(3), 135–40. (doi:10.1007/BF00192284).

Phogat, V., Skewes, M. A., Cox, J. W., Sanderson, G., Alam, J. and Šimůnek, J. 2014. Seasonal simulation of water, salinity and nitrate dynamics under drip irrigated mandarin (*Citrus reticulata*) and assessing management options for drainage and nitrate leaching. *J. Hydrol.* 513, 504–16. doi:10.1016/j.jhydrol.2014.04.008.

Rahn, C. R., Zhang, K., Lillywhite, R., Ramos, C., Doltra, J., De Paz, J. M., Riley, H., Fink, M., Nendel, C., Thorup-Kristensen, K. and Pedersen, A. 2010. EU-Rotate_N-a decision support system to predict environmental and economic consequences of the management of nitrogen fertiliser in crop rotations. *Eur. J. Hortic. Sci.* 75(1), 20–32.

Rawlins, S. L. and Raats, P. A. C. 1975. Prospects for high-frequency irrigation. *Science* 188(4188), 604–10. doi:10.1126/science.188.4188.604.

Savvas, D., Karagianni, V., Kotsiras, A., Demopoulus, V., Karkamisi, I. and Pakou, P. 2003. Interactions between ammonium and pH of the nutrient solution supplied to gerbera (*Gerbera jamesonii*) grown in pumice. *Plant Soil* 254(2), 393–402. doi:10.1023/A:1025595201676.

Segal, E., Ben-Gal, A. and Shani, U. 2006. Root water uptake efficiency under ultra-high irrigation frequency. *Plant Soil* 282(1–2), 333–41. (doi:10.1007/s11104-006-0003-6).

Silber, A., Ganmore-Neumann, R. and Ben-Jaacov, J. 1998. Effects of nutrient addition on growth and rhizosphere pH of Leucadendron 'Safari Sunset'. *Plant Soil* 199, 205–11.

Silber, A., Xu, G., Levkovitch, I., Soriano, S., Bilu, A. and Wallach, R. 2003. High fertigation frequency: the effects on uptake of nutrients, water and plant growth. *Plant Soil* 253(2), 467–77. doi:10.1023/A:1024857814743.

Simunek, J., Sejna, M. and van Genuchten, M. T. 1999. The HYDRUS-2D software package for simulating two-dimensional movement of water, heat, and multiple solutes in variably saturated media, Version 2.0, Rep. IGWMCTPS-53. IGWMC, Colorado School of Mines, Golden, CO, 251p.

Sonneveld, C. and De Kreij, C. 1999. Response of cucumber (*Cucumis sativus L.*) to an unequal distribution of salts in the root environment. *Plant Soil* 209(1), 47–56.

Sonneveld, C. and Voogt, W. 1990. Response of tomato (*Lycopersicon esculentum*) to an unequal distribution of nutrients in the root environment. *Plant Soil* 124, 251–6.

Subbiah, R. 2013. A review of models for predicting soil water dynamics during trickle irrigation. *Irrig. Sci.* 31(3), 225–58. doi:10.1007/s00271-011-0309-x.

Tanaka, A., Fujita, K. and Kikuchi, K. 1974. Nutrio-physiological studies on the tomato plant. I. Outline of growth and nutrient absorption. *Soil Sci. Plant Nutr.* 20(1), 57–68. doi:10.1080/00380768.1974.10433228.

Thompson, T. L., White, S. A., Walworth, J. and Sower, G. J. 2003. Fertigation frequency for subsurface drip-irrigated broccoli. *Soil Sci. Soc. Am. J.* 67(3), 910–8. doi:10.2136/sssaj2003.0910.

Voogt, W. and Bar-Yosef, B. 2019. Water and nutrient management and crops response to nutrient solution recycling in soilless growing systems in greenhouses. In: Raviv, M., Lieth, J. H. and Bar-Tal, A. (Eds), *Soilless Culture: Theory and Practice*. Elsevier Academic Press, UK, pp. 425–508.

Voogt, W., Van Winkel, A. and Steinbuch, F. 2006. Evaluation of the fertigation model, a decision support system for water and nutrient supply for soil grown greenhouse crops. In: *III International Symposium on Models for Plant Growth, Environmental Control and Farm Management in Protected Cultivation 718*. International Society for Horticultural Science, Leuven, Belgium, pp. 531–8.

Voogt, W., Beerling, E. A. M., Blok, C., van der Maas, A. A. and Os, E. A. van. 2013. The road to sustainable water and nutrient management in soilless culture in Dutch greenhouse horticulture. In: D'Haene, K., Vandecasteele, B., De Vis, R., Crappé,

S., Callens, D., Mechant, E., Hofman, G. and De Neve, S. (Eds), *NUTRIHORT: Nutrient Management, Nutrient Legislation and Innovative Techniques in Intensive Horticulture*, 16–18 September 2013, Ghent, Belgium. Institute for Agricultural and Fisheries Research (ILVO), Merelbeke, Belgium.

Williamson, L. C., Ribrioux, S. P. C. P., Fitter, A. H. and Leyser, H. M. O. 2001. Phosphate availability regulates root system architecture in *Arabidopsis*. *Plant Physiol.* 126(2), 875–82. doi:10.1104/pp.126.2.875.

Xu, G. H., Wolf, S. and Kafkafi, U. 2001. Effect of varying nitrogen form and concentration during growing season on sweet pepper flowering and fruit yield. *J. Plant Nutr.* 24(7), 1099–116. doi:10.1081/PLN-100103806.

Xu, G., Levkovitch, I., Soriano, S., Wallach, R. and Silber, A. 2004. Integrated effect of irrigation frequency and phosphorus level on lettuce: yield, P uptake and root growth. *Plant Soil* 263(1), 297–309. doi:10.1023/B:PLSO.0000047743.19391.42.

Yasuor, H., Ben- Gal, A., Yermiyahu, U., Beit-Yannai, E. and Cohen, S. 2013. Nitrogen management of greenhouse pepper production: agronomic, nutritional, and environmental implications. *HortScience* 48(10), 1241-9. doi:10.21273/HORTSCI.48.10.1241.

Yermiyahu, U., Faingold, I., Aldanfiri, J. and Tergenman, M. 2005. The effect of nutrition on herb plants. Final Report to the Chief Scientist of the Ministry of Agriculture, Israel (in Hebrew).

Printed in the USA
CPSIA information can be obtained
at www.ICGtesting.com
JSHW012254170624
64962JS00006B/56

9 781801 466516